ESP8266 Robotics Projects

DIY Wi-Fi controlled robots

Pradeeka Seneviratne

BIRMINGHAM - MUMBAI

ESP8266 Robotics Projects

First published: November 2017

Production reference: 1281117

Published by Packt Publishing Ltd.
Livery Place
35 Livery Street
Birmingham
B3 2PB, UK.

ISBN 978-1-78847-461-0

www.packtpub.com

Credits

Author
Pradeeka Seneviratne

Reviewer
Shweta Gupte

Commissioning Editor
Gebin George

Acquisition Editor
Prachi Bisht

Content Development Editor
Trusha Shriyan

Technical Editor
Sayali Thanekar

Copy Editors
Safis Editing
Ulka Manjrekar

Project Coordinator
Kinjal Bari

Proofreader
Safis Editing

Indexer
Pratik Shirodkar

Graphics
Kirk D'Penha

Production Coordinator
Nilesh Mohite

About the Author

Pradeeka Seneviratne is a software engineer with over 10 years of experience in computer programming and systems design. He is an expert in the development of Arduino and Raspberry Pi-based embedded systems. Pradeeka is currently a full-time embedded software engineer who works with embedded systems and highly scalable technologies. Previously, he worked as a software engineer for several IT infrastructure and technology servicing companies. He collaborated with the Outernet project as a volunteer hardware and software tester for Lighthouse and Raspberry Pi-based DIY Outernet receivers based on Ku band satellite frequencies. He is also the author of five books:

- *Internet of Things with Arduino Blueprints* [Packt Publishing]
- *IoT: Building Arduino-Based Projects* [Packt Publishing]
- *Building Arduino PLCs* [Apress]
- *Raspberry Pi 3 Projects for Java Programmers* [Packt Publishing]
- *Beginning BBC micro:bit* [Apress]

About the Reviewer

Shweta Gupte is an embedded systems engineer with 4 years, experience in the robotics field. Shweta has worked with robots both at the software level and with a low-level hardware interface. She has experience of working with embedded system design, RTOS, microcontroller programming, and the low-level interfacing of robots, motion planning, and computer vision.

Shweta has experience in programming in C, Embedded C, JAVA, MATLAB, and microcontroller assembly language. She has worked with low-level controllers, including Renesas RX62, Renesas QSK, Arduino, PIC, and Rasberry Pi.

Currently, she is working at Mathworks to develop MATLAB and Simulink products for robotics and embedded targets, including developing automated test beds and hardware testing. She has a masters degree in electrical engineering and a bachelors degree in electronics and telecommunications.

Her interests include playing with low-cost microcontroller-based robots and quadrotors. She likes to spend her spare time rock climbing and fostering rescued dogs.

www.PacktPub.com

For support files and downloads related to your book, please visit www.PacktPub.com.

Did you know that Packt offers eBook versions of every book published, with PDF and ePub files available? You can upgrade to the eBook version at www.PacktPub.com and as a print book customer, you are entitled to a discount on the eBook copy. Get in touch with us at service@packtpub.com for more details.

At www.PacktPub.com, you can also read a collection of free technical articles, sign up for a range of free newsletters and receive exclusive discounts and offers on Packt books and eBooks.

https://www.packtpub.com/mapt

Get the most in-demand software skills with Mapt. Mapt gives you full access to all Packt books and video courses, as well as industry-leading tools to help you plan your personal development and advance your career.

Why subscribe?

- Fully searchable across every book published by Packt
- Copy and paste, print, and bookmark content
- On demand and accessible via a web browser

Customer Feedback

Thanks for purchasing this Packt book. At Packt, quality is at the heart of our editorial process. To help us improve, please leave us an honest review on this book's Amazon page at `http://www.amazon.com/dp/1788474619`.

If you'd like to join our team of regular reviewers, you can email us at `customerreviews@packtpub.com`. We award our regular reviewers with free eBooks and videos in exchange for their valuable feedback. Help us be relentless in improving our products!

Table of Contents

Preface

This book is all about robotics projects based on the original ESP8266 microcontroller board and some variants of ESP8266 boards. The ESP8266 Wi-Fi module is a self-contained SOC with an integrated TCP/IP protocol stack that can give any microcontroller access to your Wi-Fi network. It has powerful processing and storage capabilities. It also supports application hosting and Wi-Fi networking.

The first chapter explains everything that you need to build your development environment with basic hardware and software components. This book uses an original ESP8266 board and an AdafruitFeather HUZZAH ESP8266 board for all the robotic projects. You will also learn how to use different types of chassis kits, motors, motor drivers, power supplies and distribution boards, sensors, and actuators to build robotic projects that can be controlled via Wi-Fi. You will also use line sensors, ArduCam, wheel encoders, and a Gripper Kit to build more specialized robots.

What this book covers

Chapter 1, *Getting Ready*, introduces the original ESP8266, that is, ESP-01, sets up a development environment with the hardware and software needed to write code for ESP8266, and explains how to use the Arduino Core to write code for ESP8266 boards with the Arduino IDE. This chapter will also present how to use the AT command with PuTTY to execute on an ESP8266.

Chapter 2, *Building a Mini Round Robot with Original ESP8266*, explains how to build a Mini Round Robot using the Mini 3-Layer Round Robot Chassis Kit with the ESP-01 board, Arduino, and some electronics. This chapter will also explain how to develop a Blynk app to control the robot through a Wi-Fi network.

Chapter 3, *Using Encoders*, shows how to modify the Mini 3-Layer Round Robot you built in the previous chapter by integrating a wheel encoder kit to calculate average speed and distance traveled based on the pulses generated by the hall-effect sensors. This chapter will also explain how to modify the Blynk app you built in the previous chapter to show the calculated average speed and distance traveled.

Chapter 4, *Building a Mini Round Robot with the Feather HUZZAH ESP8266*, shows how to replace the original ESP8266 board with a physically smaller and more lightweight Feather HUZZAH ESP8266 board. The chapter will also explain how to develop a Blynk app to control the robot through a Wi-Fi network.

Chapter 5, *Line-Following Zumo Robot*, teaches how to build a robot that follows a line, either a black line on a white surface or a white line on a black surface, by integrating a Line Follower Array. The chapter will also explain how to build a line following course.

Chapter 6, *Building an ESP8266 Robot Controller*, shows how to build an ESP8266-based simple Robot Controller that can be used to control a Romi Robot through a Wi-Fi network using Blynk Bridge.

Chapter 7, *Building a Gripper Robot*, explains how to build a robot based on the Mini Robot Rover chassis, a parallel Gripper Kit, and a servo motor. The chapter will also explain how to build a Blynk app to control the gripper through a Wi-Fi network.

Chapter 8, *Photo Rover Robot*, shows how to build a rover robot that can be used to take pictures with an ArduCAM from remote locations and view them using a web browser. The robot uses a web socket server that allows you to control the camera using a simple web-based interface. The same interface can be used to view the captured image as well.

What you need for this book

To get the most from this book, you need a basic knowledge of programming with the Arduino IDE and electronics.

Who this book is for

This book is for those who are familiar with robotics and want to build robotics projects using the ESP8266 microcontroller.

Conventions

In this book, you will find a number of text styles that distinguish between different kinds of information. Here are some examples of these styles and an explanation of their meaning.

Code words in text, database table names, folder names, filenames, file extensions, pathnames, dummy URLs, user input, and Twitter handles are shown as follows: "Copy all these tools to `your_sketchbook_folder/tools` of the Arduino IDE."

A block of code is set as follows:

```
#include BLYNK_WRITE(V1)
{
intx = param[0].asInt(); // assigning incoming value from pin V1 to a
variable
inty = param[1].asInt(); // assigning incoming value from pin V1 to a
variable
```

New terms and **important words** are shown in bold. Words that you see on the screen, for example, in menus or dialog boxes, appear in the text like this: "Run the mini round robot project in the Blynk app by tapping on the **Play** button."

Warnings or important notes appear like this.

Tips and tricks appear like this.

Reader feedback

Feedback from our readers is always welcome. Let us know what you think about this book—what you liked or disliked. Reader feedback is important for us as it helps us develop titles that you will really get the most out of.

To send us general feedback, simply email feedback@packtpub.com, and mention the book's title in the subject of your message.

If there is a topic that you have expertise in and you are interested in either writing or contributing to a book, see our author guide at www.packtpub.com/authors.

Customer support

Now that you are the proud owner of a Packt book, we have a number of things to help you to get the most from your purchase.

Downloading the example code

You can download the example code files for this book from your account at `http://www.packtpub.com`. If you purchased this book elsewhere, you can visit `http://www.packtpub.com/support` and register to have the files emailed directly to you.

You can download the code files by following these steps:

1. Log in or register on our website using your email address and password.
2. Hover the mouse pointer on the **SUPPORT** tab at the top.
3. Click on **Code Downloads & Errata**.
4. Enter the name of the book in the **Search** box.
5. Select the book for which you're looking to download the code files.
6. Choose from the drop-down menu where you purchased this book from.
7. Click on **Code Download**.

Once the file is downloaded, please make sure that you unzip or extract the folder using the latest version of:

- WinRAR / 7-Zip for Windows
- Zipeg / iZip / UnRarX for Mac
- 7-Zip / PeaZip for Linux

The code bundle for the book is also hosted on GitHub at `https://github.com/PacktPublishing/ESP8266-Robotics-Projects`. We also have other code bundles from our rich catalog of books and videos available at `https://github.com/PacktPublishing/`. Check them out!

Downloading the color images of this book

We also provide you with a PDF file that has color images of the screenshots/diagrams used in this book. The color images will help you better understand the changes in the output. You can download this file from `https://www.packtpub.com/sites/default/files/downloads/ESP8266RoboticsProjects_ColorImages.pdf`.

Errata

Although we have taken every care to ensure the accuracy of our content, mistakes do happen. If you find a mistake in one of our books-maybe a mistake in the text or the code-we would be grateful if you could report this to us. By doing so, you can save other readers from frustration and help us improve subsequent versions of this book. If you find any errata, please report them by visiting `http://www.packtpub.com/submit-errata`, selecting your book, clicking on the **Errata Submission Form** link, and entering the details of your errata. Once your errata are verified, your submission will be accepted and the errata will be uploaded to our website or added to any list of existing errata under the Errata section of that title. To view the previously submitted errata, go to `https://www.packtpub.com/books/content/support` and enter the name of the book in the search field. The required information will appear under the **Errata** section.

Piracy

Piracy of copyrighted material on the internet is an ongoing problem across all media. At Packt, we take the protection of our copyright and licenses very seriously. If you come across any illegal copies of our works in any form on the internet, please provide us with the location address or website name immediately so that we can pursue a remedy.

Please contact us at `copyright@packtpub.com` with a link to the suspected pirated material.

We appreciate your help in protecting our authors and our ability to bring you valuable content.

Questions

If you have a problem with any aspect of this book, you can contact us at `questions@packtpub.com`, and we will do our best to address the problem.

1
Getting Ready

Welcome to the exciting world of programming **ESP8266** with **Arduino core for ESP8266**. This chapter introduces ESP8266 and a simple LED blink program using the Arduino core for the ESP8266 library.

In this chapter, you will learn the following topics:

- The basics of ESP8266 and the module family
- The original ESP8266 ESP-01
- Applying regulated power to ESP-01
- Connecting ESP-01 with computer through a USB to TTL serial console cable
- Preparing the development environment by installing the Arduino core for the ESP8266 library
- Writing a simple LED blink program using the Arduino core for ESP8266
- Using AT commands
- The overview of some robotics hardware and accessories

ESP8266EX

The ESP8266EX (*Figure 1.1*) is a single-chip microcontroller developed by **Espressif** (https://espressif.com/). It can be used to develop low-power, highly-integrated Wi-Fi solutions especially for mobile devices, wearable electronics, and the **Internet of Things (IoT)** applications. The main advantage of ESP8266EX is that it can give any microcontroller access to a Wi-Fi network. As an example, it can offload all Wi-Fi networking functions for Arduino just like the Arduino Wi-Fi shield:

Figure 1.1: ESP8266EX chip

The chip has the following great features:

- Small size (5 mm x 5 mm)
- Requires minimal external circuitry
- Wide temperature range (-40°C to + 125°C)
- All-in-one small package -SoC (integrates a 32-bit Tensilica MCU, standard digital peripheral interfaces, antenna switches, RF balun, power amplifier, low-noise receive amplifier, filters, and power management modules)
- Low power consumption

ESP-01

ESP-01 is a small Wi-Fi module based on **Espressif ESP8266EX** that can give any microcontroller access to a Wi-Fi network. The ESP-01 module (*Figure 1.2*) is the basic and the most popular board of the ESP8266 module family. You should have an ESP8266 ESP-01 module to understand the basics of ESP8266 and build robotics projects.

This is a list of vendors that typically stock the ESP-01 modules:

- **SparkFun Electronics**: https://www.sparkfun.com/products/13678
- **Adafruit Industries**: https://www.adafruit.com/product/2282
- **Electrodragon**: http://www.electrodragon.com/product/esp8266-wi07c-wifi-module/

Figure 1.2: The original ESP8266 ESP-01. Image courtesy of SparkFun Electronics (https://www.sparkfun.com)

The original ESP8266 ESP-01 has the following features:

- Offloads all Wi-Fi networking functions from another application's processor
- Self-hosting of applications
- Only four pins interface, VCC-3V3, GND, TX, and RX
- 1 MB (8 MB) flash memory
- Pre-programmed AT command set firmware
- Supports APSD for VoIP applications and Bluetooth co-existence interfaces

Here is a list of all the ESP8266 boards of the ESP8266 module family at the time of writing:

- ESP-01
- ESP-02
- ESP-03
- ESP-04
- ESP-05
- ESP-06
- ESP-07
- ESP-08
- ESP-09
- ESP-10
- ESP-11
- ESP-12
- ESP-12F
- ESP-12-E/Q
- ESP-12S
- ESP-13
- ESP-14

Board features and connections

The top of the ESP-01 board consist of the following things:

- An ESP8266EX chip—Offloading all Wi-Fi networking functions
- A Berg Micro BG25Q80A 1518 flash chip—**Serial Peripheral Interface** (**SPI**) flash memory
- A printed Wi-Fi antenna—Transmitting and receiving Wi-Fi signals

Figure 1.3: ESP-01 top view

Connections

The original ESP8266 ESP-01 board has eight pins and comes without headers. You can solder regular wires with pins, but if you want to use it with a breadboard, you should attach headers:

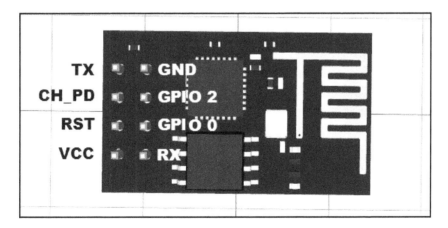

Figure 1.4: ESP-01 pinout

- **GPIO0**: Connects to pin 3 of the ESP8266
- **GPIO2**: Connects to pin 5 of the ESP8266
- **TX**: This is the transmitting line that connects to RX of another device
- **RX**: This is the receiving line that connects to TX of another device
- **3V**: This accepts regulated 3.3V DC from a regulator to power the board
- **GND**: The ground pin
- **CH_PD**: Chip select
- **RST**: The reset pin

Connecting with a breadboard

You can use one of the following things to connect ESP8266 with a breadboard:

- Using female/male extension jumper wires (`https://www.adafruit.com/product/1954`)
- Using a 10-pin IDC breakout helper (`https://www.adafruit.com/product/2102`)

Power supply

The ESP8266 works only with regulated 3.3V. You can build a simple regulated power supply with an LD1117-3.3 linear voltage regulator. It can provide 3.3V at 800 mA. You can purchase a TO-220 package of the LD1117-3.3 for use with a breadboard. *Figure 1.5* shows the circuit diagram for 3.3V output. You can supply 4-15V DC to V_{in} using one of the following sources:

- 7V LiPo battery
- 5V power bank
- 9V battery
- 5V "wall wart" power supply

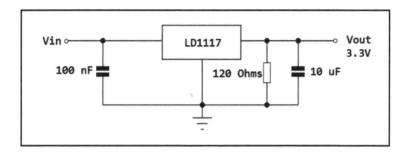

Figure 1.5: 3.3V regulated power supply circuit diagram

Now, you can connect the power supply to ESP-01:

1. Connect V_{OUT} to the V_{cc} pin of ESP-01.
2. Connect GND to the GND pin of ESP-01.

Connecting through USB for flashing

ESP8266 ESP-01 doesn't have a built-in USB to serial conversion chip to directly connect to your computer through a USB cable for flashing programs. As a solution, you can use a USB to TTL serial console cable to connect ESP8266 to a computer. *Figure 1.6* shows a USB TTL serial console cable from Adafruit (https://www.adafruit.com/product/954):

Figure 1.6: A USB to TTL serial console cable. Image courtesy of Adafruit Industries (https://www.adafruit.com)

The type A USB plug includes a USB to serial conversion chip and provides four wires to connect to your ESP8266ESP-01 board. Table 1.1 shows the functions of the four wires:

Color	Function
Red	Power
Black	Ground
White	RX
Green	TX

Table 1.1: Color codes and functions of wires

You can use the following steps to connect the USB to TTL serial cable with your ESP8266:

1. Connect the white RX wire of the cable to the TX pin of ESP8266
2. Connect the green TX wire of the cable to the RX pin of ESP8266
3. Connect the black ground wire of the cable to the GND pin of ESP8266
4. Connect the CH_PD pin of ESP8266 to the V_{cc} wire of the power supply

Don't connect the red power wire because it is 5V.

Using a serial terminal program

The following steps will guide you how to use a serial terminal program to communicate and execute AT commands with ESP-01. In this example, you will use PuTTY, SSH, and a Telnet client for Windows to run AT commands through a serial port:

1. First, download the latest PuTTY MSI installer or binary from the following sources:
   - MSI (Windows installer):
     - **32-bit**: `putty-0.70-installer.msi` (`https://the.earth.li/~sgtatham/putty/latest/w32/putty-0.70-installer.msi`)
     - **64-bit**: `putty-64bit-0.70-installer.msi` (`https://the.earth.li/~sgtatham/putty/latest/w64/putty-64bit-0.70-installer.msi`)
   - Binary:
     - **32-bit**: `putty.exe` (`https://the.earth.li/~sgtatham/putty/latest/w32/putty.exe`)
     - **64-bit**: `putty.exe` (`https://the.earth.li/~sgtatham/putty/latest/w64/putty.exe`)

2. Open the **PuTTY Configuration** window by double clicking on `putty.exe` or the putty shortcut.
3. In the **Serial line** text box, type the COM port that is assigned to your ESP8266 board (*Figure 1.7*).
4. In the **Speed** text box, type `115200` as the baud rate.

5. Click on the **Open** button to make a serial connection:

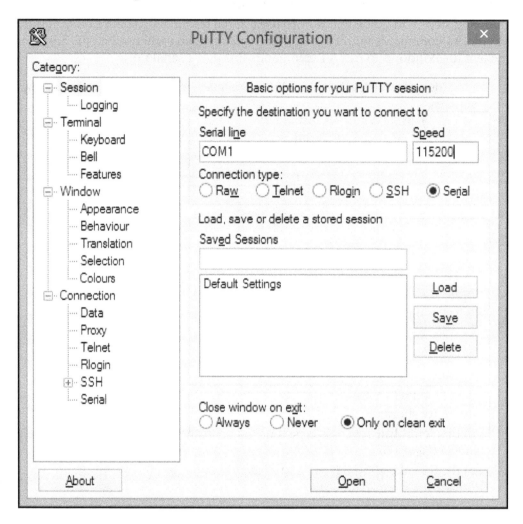

Figure 1.7: The PuTTY configuration window

6. If you have provided the correct settings, the serial connection will establish between ESP8266 and your computer.
7. You can use the serial terminal window to issue AT commands for ESP8266 from your computer.

AT commands

Table 1.2 shows the complete set of AT commands that you can use with ESP8266:

AT command	Manual
AT	Attention.
AT+RST	Resetting the unit.
AT+GMR	Retrieving the firmware version ID.
AT+CWMODE=? AT+CWMODE? AT+CWMODE=<mode>	Setting operation mode: • Client • Access point • Client and access point The access point functionality does not have a DHCP function and has only minimum functionalities. However, it will assign an IP address to the client and there is no way to do a manual IP, manual DNS, and other advanced IP functionalities. This unit only provides minimal functionalities.
AT+CWJAP=<ssid>,<pwd> AT+CWJAP?	Joining a network or just an access point.
AT+CWLAP	Retrieving the list of the visible network.
AT+CWQAP	Disconnecting from the current network connection.

`AT+CWSAP=<ssid>,<pwd>,<chi>,<ecn>` `AT+CWSAP?`	Setting up access point SSID, password, RF channel, and security scheme. The following is the security scheme: • `0`: Open. No security. • `1`: `WEP`. • `2`: `WPA_PSK`. • `3`: `WPA2_PSK`. • `4`: `WPA_WPA2_PSK`.
`AT+CWLF`	Retrieving a list of assigned IP addresses.
`AT+CIPSTATUS`	Retrieving the current connection as socket client or socket server.

`AT+CIPSTART=?`
`AT+CIPSTART=<type>,<addr>,<port>`
`(AT+CIPMUX=0)`
`AT+CIPSTART=<id>,<type>,<port>`
`(AT+CIPMUX=1)`
	Connecting to socket server (TCP or UDP).

`AT+CIPSEND=<length> (AT+CIPMUX=0 &`
`AT+CIPMODE=0)`
`AT+CIPSEND=<id>,<length> (AT+CIPMUX=1`
`& AT+CIPMODE=0)`
`AT+CIPSEND (AT+CIPMUX=0 &`
`AT+CIPMODE=1)`
	Sending by connection channel and by specific length.

`AT+CIPCLOSE`	Closing the socket connection.
`AT+CIFSR`	Retrieving the assigned IP address when the unit is connecting to a network.
`AT+ CIPMUX=AT+CIPMUX?`	Setting a single connection (`AT+CIPMUX=0`) or multi-channel connection (`AT+CIPMUX=1`).
`AT+CIPSERVER= [,] (AT+CIPMUX=1)`	Starting at the specified port or stopping the server. The default port is `333`. `<mod>` is as follows: • `0`: Close the socket server • `1`: Open the socket server

`AT+CIPMODE=<mode>` `AT+CIPMODE?`	Setting transparent mode (data from the socket client will be sent to the serial port as is) or connection channel specific mode (`+IPD,<connection channel>,<length>`) segments. Data sent from the socket client will be broken into multiple unsolicited (`+IPD,<connection channel>,<length>`) segments. `<mode>`is as follows: • `0`: Data received will be sent to the serial port with a `+IPD,<connection channel>,<length>` format. (`AT+CIPMUX=[0,1]`) • `1`: Data received will be sent to the serial port as a data stream. (`AT+CIPMUX=0`)
`AT+CIPSTO=<time>` `AT+CIPSTO?`	Setting the automatic socket client disconnection timeout from `1` to `28800` seconds due to inactivities.
Packetized data from the unit	Unsolicited data packet (`+IPD, <connection channel>,<length>`).

Table 1.2: The ESP8266 AT command set (Source: SparkFun Electronics)

Using AT commands

Every AT command starts with the letters AT. Here's an example:

AT+GMR <CR>

Type the AT command AT+GMR and press the *Enter* key. Don't type <CR>. The CR indicates carriage return.

This will return the firmware version ID of the ESP8266 module, as shown in *Figure 1.8*:

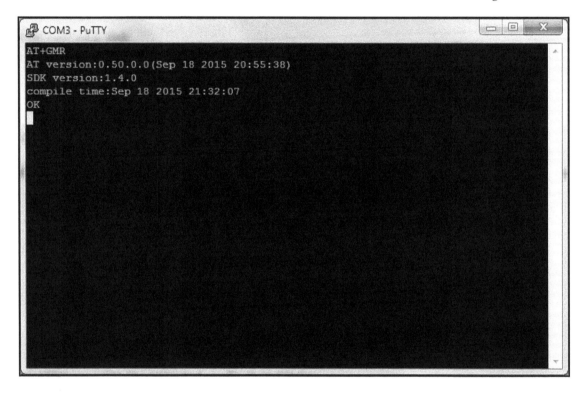

Figure 1.8: Terminal output for the AT+GMR command

Likewise, you can run any AT command with the PuTTY terminal to execute on ESP8266 through the serial connection.

Using the Arduino IDE

The Arduino IDE provides an easy way to write sketches for Arduino and other development boards. You can use ESP8266 with the Arduino IDE. First, install the ESP8266 board package on Arduino IDE.

The ESP8266 also supports the following development environments:

- Use a simple Notepad/GCC setup
- Fine-tune an Eclipse environment
- Use a virtual machine provided by Espressif

Installing the Arduino core for an ESP8266 Wi-Fi chip

The Arduino core for ESP8266 includes a set of libraries to do the following things with your ESP8266 Wi-Fi module:

- Communicate over Wi-Fi using TCP and UDP protocols
- Set up HTTP, mDNS, SSDP, and DNS servers
- Perform OTA updates
- Use a filesystem in flash memory
- Work with SD cards, servos, SPI, and I2C peripherals

It also allows you to write sketches for ESP8266 with the core Arduino functions and libraries. You can run the sketches directly on ESP8266 without using an external microcontroller board:

1. Open the Arduino IDE and on the menu bar select **File** | **Preferences**.

2. Type the URL
   `http://arduino.esp8266.com/stable/package_esp8266com_index.json`
   for the ESP8266 board manager package for Arduino in the **Additional Boards Manager URLs** text box and click on **OK** to close the dialog box (*Figure 1.9*):

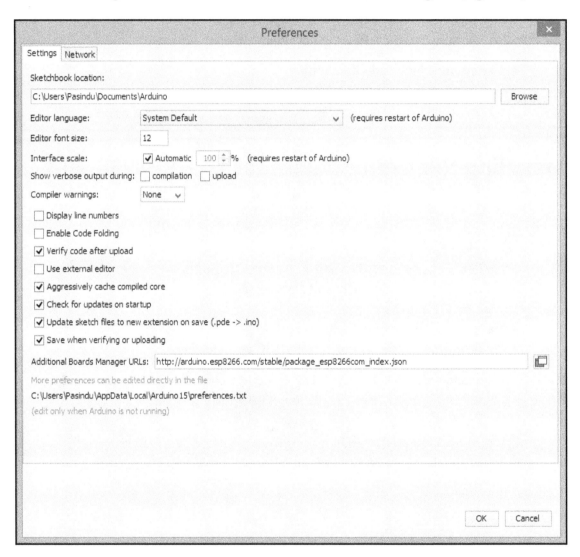

Figure 1.9: The Arduino IDE preferences

3. Open the **Boards Manager** by selecting **Tools | Board | Boards Manager**. You should see a new entry in the list titled **esp8266 by ESP8266 Community** (*Figure 1.10*):

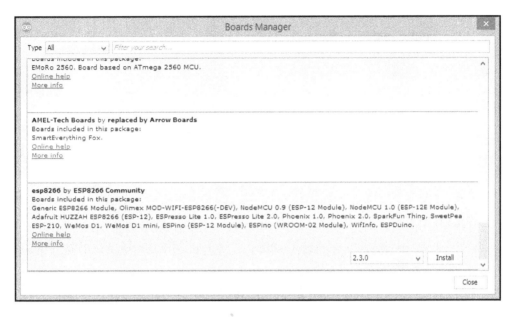

Figure 1.10: The Arduino IDE Boards Manager

4. At the time of this writing, the Arduino core for ESP8266 Wi-Fi chip supports the following boards:
   - Generic ESP8266 Module
   - Olimex MOD-WIFI-ESP8266(-DEV)
   - NodeMCU 0.9 (ESP-12 Module)
   - NodeMCU 1.0 (ESP-12E Module)
   - Adafruit HUZZAH ESP8266 (ESP-12)
   - ESPresso Lite 1.0
   - ESPresso Lite 2.0
   - Phoenix 1.0
   - Phoenix 2.0
   - SparkFun Thing
   - SweetPea ESP-210
   - WeMos D1
   - WeMos D1 mini

- ESPino (ESP-12 Module)
- ESPino (WROOM-02 Module)
- WifInfo
- ESPDuino

5. Click on the **Install** button to install it on your Arduino IDE. This will take a few minutes to install, depending on your internet speed.
6. Click on **Close** to close the **Boards Manager** dialog box.
7. Select **Tools** | **Board: "Generic ESP8266 Module"** | **Generic ESP8266** for board type.
8. The upload speed should be 115200; however, you can increase it to a higher value by clicking on **Tools** | **Upload Speed** and selecting a value greater than 115200.

Hello world

Now, you will write your first program (sketch) for ESP8266 with the Arduino IDE to blink a LED connected to PIN 2 of ESP8266.

You'll need the following things to build the circuit:

- One LED (any color)
- One 1 kilo ohm resistor
- 3V regulated power supply

Connect the components together, as shown in *Figure 1.11*:

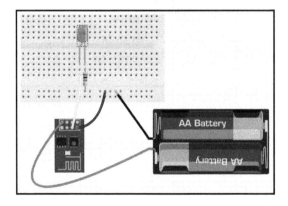

Figure 1.11: The LED blink circuit hook up

Listing 1-1 – Blink a LED

Using your Arduino IDE, type the code as shown in the *Listing 1-1*:

```
#define ESP8266_LED 2
void setup()
{
  pinMode(ESP8266_LED, OUTPUT);
}
void loop()
{
  digitalWrite(ESP8266_LED, HIGH);
  delay(500);
  digitalWrite(ESP8266_LED, LOW);
  delay(500);
}
```

Then, verify the sketch by clicking on the **V**erify button. Finally, upload the program by clicking on the **Upload** button. This will take a few seconds to complete. After completing the flashing, the LED will start to blink.

Using chassis kits and accessories to build robots

Chassis kits provide everything to build robots with most microcontroller boards. You'll use the following chassis kits to build projects.

Mini 3-Layer Round Robot Chassis Kit

The 3-Layer Mini Round Robot chassis Kit (*Figure 1.12*) includes the following things you need to build the shell of a two-wheel drive mobile platform robot:

- Two drive motors
- Two wheels
- One plastic caster ball
- Anodized aluminum frames and all mounting hardware for assembly

You can purchase a kit from Adafruit Industries (`https://www.adafruit.com/product/3244`) to work with projects that will be discussed in `Chapter 2`, *Building a Mini Round Robot with Original ESP8266*, `Chapter 3`, *Using Encoders*, and `Chapter 4`, *Building a Mini Round Robot with the Feather HUZZAH ESP8266*:

Figure 1.12: A Mini Round Robot Chassis Kit. Image courtesy of Adafruit Industries (https://www.adafruit.com)

Zumo chassis kit

The Zumo chassis kit (*Figure 1.13*) allows you to build tracked robots that run on continuous tracks instead of wheels. Tracked vehicles can be used for:

- Construction vehicles
- Military armored vehicles
- Unmanned ground vehicles

The Zumo chassis kit includes the following things:

- Zumo chassis main body
- 1/16" black acrylic mounting plate
- Two drive sprockets
- Two idler sprockets
- Two 22-tooth silicone tracks
- Two shoulder bolts with washers and M3 nuts
- Four 1/4" #2-56 screws and nuts
- Battery terminals

You can purchase a kit from Pololu (`https://www.pololu.com/product/1418`) to work with the project that will be discussed in `Chapter 5`, *Line-Following Zumo Robot*:

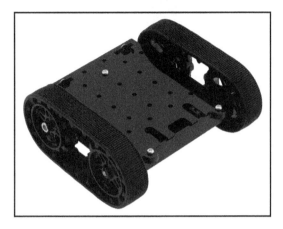

Figure 1.13: A Zumo chassis kit. Image courtesy of Pololu (https://www.pololu.com)

Romi chassis kit

The Romi chassis kit (*Figure 1.14*) includes the following things you need to build a two-wheel drive robot:

- One black Romi chassis base plate
- Two mini plastic gearmotors (120:1 HP with offset output and extended motor shaft)
- A pair of black Romi chassis motor clips
- A pair of white 70 × 8 mm Pololu Wheels
- One black Romi chassis ball caster kit
- One Romi chassis battery contact set

You can purchase a kit from Pololu (`https://www.pololu.com/category/202/romi-chassis-and-accessories`) to build the project that will be discussed in `Chapter 6`, *Building an ESP8266 Robot Controller*:

Figure 1.14: A Romi chassis kit - Black. Image courtesy of Pololu (https://www.pololu.com)

Mini robot rover chassis kit

The mini robot rover chassis kit (*Figure 1.15*) includes the following things needed to build the shell of a two-wheel drive robot:

- Two wheels
- Two DC motors in MicroServo shape
- One support wheel
- One metal chassis
- One top metal plate with mounting hardware

Figure 1.15: A mini robot rover chassis kit. Image courtesy of Adafruit Industries (https://www.adafruit.com)

You can purchase a kit from Adafruit industries
(`https://www.adafruit.com/product/2939`) to build the project that will be discussed in
`Chapter 7`, *Building a Gripper Robot*.

Rover 5 robot platform

The Rover 5 robot platform (*Figure 1.16*) allows you to build a four-wheel drive tracked
robot. The kit includes the following things:

- Adjustable gearbox angles
- 4 independent DC motors
- 4 independent Hall-effect encoders
- Thick rubber tank treads
- 6 x AA battery holder
- 10 Kg/cm stall torque per motor

Figure 1.16: A Rover 5 robot platform. Image courtesy of Pololu (https://www.pololu.com)

You can purchase a kit from Pololu (`https://www.pololu.com/product/1551`) to build a project that will be discussed in `Chapter 8`, *Photo Rover Robot*.

Wheel encoder kit

The wheel encoder kit (*Figure 1.17*) allows you to measure the speed or distance that the chassis travels. You'll learn how to connect the wheel encoder kit to any wheeled robot in `Chapter 3`, *Using Encoders*:

Figure 1.17: A wheel encoder kit. Image courtesy of SparkFun Electronics (https://www.sparkfun.com)

Parallel Gripper Kit A - Channel mount

A gripper can be connected to a robot to allow grip objects. *Figure 1.18* shows a Parallel Gripper Kit assembled. You will need any standard size Hitec or Futaba servo to integrate with the gripper kit. You can purchase one from SparkFun Electronics (`https://www.sparkfun.com/products/13178`):

Figure 1.18: Parallel Gripper Kit A - Channel mount

Summary

In this chapter, you learned the basics of the ESP8266 microcontroller and the modules based on it. Then, you learned how to run AT commands on ESP8266 and used the Arduino core for ESP8266 to write sketches. Finally, you gained knowledge on various robot chassis kits and accessories that can be used to build robots.

Chapter 2, *Building a Mini Round Robot with Original ESP8266,* offers the fundamentals of robotic vehicles by building a mini round robot with ESP-01.

2

Building a Mini Round Robot with Original ESP8266

In this chapter, you will learn how to build a two-wheeled mini round robot with the ESP8266 ESP-01 board. As you know, the nifty ESP8266 ESP01 module can handle all the necessary things related to Wi-Fi networking. The Arduino works as the main microcomputer board and the ESP01 works as a Wi-Fi shield for Arduino.

In this chapter, you will learn the following topics:

- Assembling a Mini 3-Layer Round Robot chassis kit
- Connecting ESP8266 ESP-01 with Arduino
- Using logic level converter between ESP8266 ESP-01 and Arduino
- Connecting motors with motor driver
- Connecting motor driver board with Arduino
- Power supply with battery packs (9V and 4xAA)
- Building/writing a control app with Blynk
- Moving and turning
- Controlling the robot via Wi-Fi with the Blynk app

Things you will need

First, you should prepare with the following things to build the Wi-Fi controlled mini round robot:

- 1 x Mini 3-Layer Round Robot Chassis Kit (https://www.adafruit.com/product/3244)
- 1 x ESP8266 ESP-01 board (https://www.sparkfun.com/products/13678)
- 1 x Arduino UNO - R3 (https://www.sparkfun.com/products/13678)
- 1 x SparkFun Logic Level Converter - Bi-Directional (https://www.sparkfun.com/products/12009)
- 1 x SparkFun Motor Driver - Dual TB6612FNG (1 A) (https://www.sparkfun.com/products/9457)
- 1 x Half-size breadboard (https://www.adafruit.com/product/64)
- Few hookup wires
- 1 x 9V battery holder with switch and 5.5mm/2.1mm plug (https://www.adafruit.com/product/67)
- 1 x Alkaline 9V Battery (https://www.adafruit.com/product/1321)
- 1 x 4 x AA battery holder with an on/off switch (https://www.adafruit.com/product/830)
- 1 x Alkaline AA batteries (LR6) - 4 pack (https://www.adafruit.com/product/3349)
- 1 x USB A-to-B cable
- 1 x Pocket screwdriver (https://www.adafruit.com/product/3284)

After collecting all the required things, you can start to build the first robot project step by step.

Assembling the Mini 3-Layer Round Robot chassis kit

The Mini 3-Layer Round Robot chassis kit (*Figure 2.1*) gives you everything you need to build a two-wheel drive robot chassis. The kit includes the following things:

- Three chassis layers
- Two drive motors (drive with 3-6V DC, 200-400 mA run, 1.5A hard stall)
- Two wheels

- Two rubber treads
- Two red panels
- Four brass standoffs
- Four long screws
- Eight small screws
- Four nuts
- Two small screws
- One nut
- One plastic caster ball

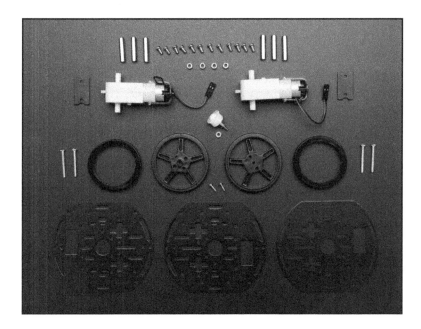

Figure 2.1: Mini 3-Layer Round Robot chassis kit. Image courtesy of Adafruit Industries (https://www.adafruit.com)

Assembling the wheels

First, assemble the two wheels with motors. Take the following things (*Figure 2.2*):

- Two motors
- Two wheels
- Four long screws
- Four nuts

- Two black panels

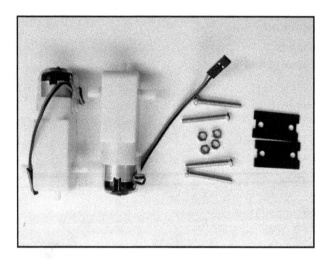

Figure 2.2: Things you will need to assemble the wheels. Image courtesy of Adafruit Industries (https://www.adafruit.com); Image credits—John O'Brien-Carelli

The following steps will guide you how to assemble them together:

1. Screw the two black panels onto the motors. The metal panels go on the side with the red and black wires coming out. Use the hex nuts to secure the screws (*Figure 2.3*):

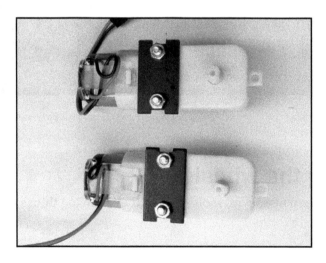

Figure 2.3: Screwing the red panels onto the motors. Image courtesy of Adafruit Industries (https://www.adafruit.com); Image credits—John O'Brien-Carelli

2. Take the two wheels, rubber treads, and two small screws (*Figure 2.4*):

Figure 2.4: Wheels and rubber threads. Image courtesy of Adafruit Industries (https://www.adafruit.com); Image credits—John O'Brien-Carelli

3. Put the rubber treads on the wheels. Then fit the wheels onto the white knob on the motors to snap onto the oval center. Finally, attach the wheels into place with tiny screws (*Figure 2.5*):

Figure 2.5: Attaching the wheel onto the motor. Image courtesy of Adafruit Industries (https://www.adafruit.com); Image credits—John O'Brien-Carelli

Assembling the chassis

The following steps will guide you how to assemble the chassis with the provided things:

1. Take one of the chassis layers. The two layers are identical. Align the chassis layer on your workbench (*Figure 2.6*). The rectangular cut should be on the left side:

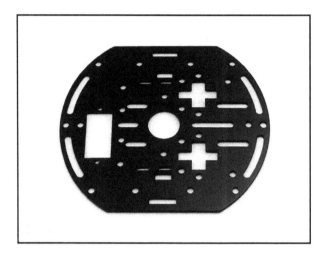

Figure 2.6: A chassis layer. Image courtesy of Adafruit Industries (https://www.adafruit.com); Image credits—John Park

2. Attach two brass standoffs onto the chassis layer (*Figure 2.7*):

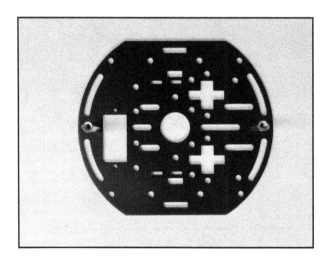

Figure 2.7: Attaching the brass standoffs. Image courtesy of Adafruit Industries (https://www.adafruit.com); Image credits—John O'Brien-Carelli

3. Turn the chassis layer and attach the caster wheel onto it, as shown in *Figure 2.8*:

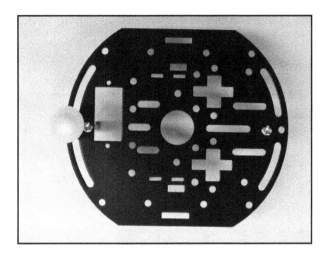

Figure 2.8: Attaching the caster wheel. Image courtesy of Adafruit Industries (https://www.adafruit.com); Image credits—John O'Brien-Carelli

4. Turn over the chassis layer again.
5. Take the assembled wheels and place them on the chassis layer (*Figure 2.9*):

Figure 2.9: Fitting assembled wheels onto the chassis layer. Image courtesy of Adafruit Industries (https://www.adafruit.com); Image credits—John O'Brien-Carelli

Attaching 9V battery box to the chassis layer

Open the 9V battery box and insert the 9V battery. Make sure to switch off the battery box before inserting the battery:

1. Apply a piece of double tape to the battery box (*Figure 2.10*):

Figure 2.10: Applying a piece of double tape. Image courtesy of Adafruit Industries (https://www.adafruit.com); Image credits—John Park

2. Then place the battery box on the chassis layer between the two motors (*Figure 2.11*):

Figure 2.11: Attaching a 9V battery box. Image courtesy of Adafruit Industries (https://www.adafruit.com); Image credits—John Park

Connecting the middle chassis

Take a chassis layer and place it onto the robot and then screw in the two brass standoffs (*Figure 2.12*):

Figure 2.12: Connecting the middle chassis. Image courtesy of Adafruit Industries (https://www.adafruit.com); Image credits—John Park

Attach the four brass standoffs to the middle chassis layer (*Figure 2.13*):

Figure 2.13: Connecting the brass standoffs. Image courtesy of Adafruit Industries (https://www.adafruit.com); Image credits—Tyler Cooper

Attaching 4 x AA battery box to the chassis layer

Take the 4 x AA battery box and insert the batteries. Make sure to slide the power switch to the off position:

1. Apply rubber bumpers to the battery box (*Figure 2.14*):

Figure 2.14: Applying the rubber bumpers. Image courtesy of Adafruit Industries (https://www.adafruit.com); Image credits—Tyler Cooper

2. Flip the battery box over and place the scrap piece of the bumper material in the middle (*Figure 2.15*):

Figure 2.15: Applying the scrap piece of the bumper material. Image courtesy of Adafruit Industries (https://www.adafruit.com); Image credits—Tyler Cooper

3. Place the battery box (*Figure 2.16*):

Figure 2.16: Placing the 4 x AA battery box. Image courtesy of Adafruit Industries (https://www.adafruit.com); Image credits—Tyler Cooper

Mounting the Arduino board

Take the Arduino UNO board and mount it on the top chassis layer with screws and nuts.

Place the top chassis layer on the battery box and apply screws to the brass standoffs.

Wiring them together

After mounting the Arduino board, you can start to wire it with the motor driver and the ESP01.

Connecting the Arduino with motor driver

You can wire each pin of the motor driver with Arduino as follows:

- Motor Driver AIN1 -> Arduino Digital Pin 2
- Motor Driver BIN1 -> Arduino Digital Pin 7
- Motor Driver AIN2 -> Arduino Digital Pin 4
- Motor Driver BIN2 -> Arduino Digital Pin 8
- Motor Driver PWMA -> Arduino Digital Pin (PWM) 5

- Motor Driver PWMB -> Arduino Digital Pin (PWM) 6
- Motor Driver STBY -> Arduino Digital Pin 9

Connecting the motors with the motor driver

You will need the SparkFun Motor Driver - Dual TB6612FNG (1A) to drive the two motors. *Figure 2.17* shows the wiring diagram that you can follow to make connections between the two motors and the motor driver:

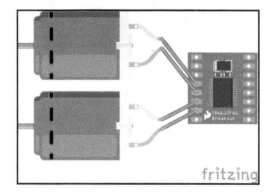

Figure 2.17: Connecting the two motors with the motor driver

1. Connect the left motor with the A01 and A02 pads of TB6612FNG.
2. Connect the right motor with the B01 and B02 pads of TB6612FNG.

Connecting the ESP01 with Arduino

The ESP01 works as a Wi-Fi shield for the Arduino UNO board. The ESP01 is directly powered by the Arduino UNO 3.3V regulated output.

You will use SoftwareSerial to establish serial communication between the ESP01 and Arduino UNO. You will use Arduino Digital Pins 2 and 3 for serial communication.

The Arduino works with 5V and the ESP01 works with 3.3V. You should use a logic level converter to convert the logic levels. The SparkFun bi-directional logic level converter (*Figure 2.18*) is a small device that safely steps down 5V signals to 3.3V and steps up 3.3V to 5V at the same time:

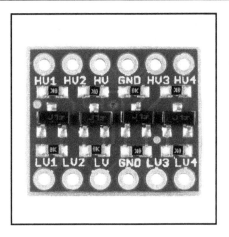

Figure 2.18: The SparkFun bi-directional logic level converter. Image courtesy of SparkFun Electronics (https://www.sparkfun.com/)

Figure 2.19 shows the wiring diagram between the ESP01 and Arduino UNO through the SparkFun bi-directional logic level converter:

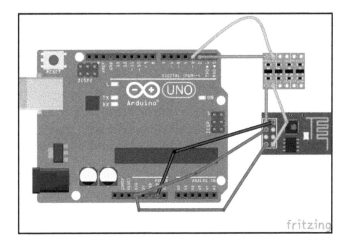

Figure 2.19: Wiring between the ESP01 and Arduino UNO through the logic level converter

Perform the following steps to wire the connections between the ESP01, Arduino, and the bi-directional logic level converter:

1. Connect ESP01 TX with the bi-directional logic level converter LV1.
2. Connect the bi-directional logic level converter HV1 with Arduino RX.
3. Connect ESP01 GND with the bi-directional logic level converter GND (low voltage side).

4. Connect ESP01 RX with the bi-directional logic level converter LV2.
5. Connect the bi-directional logic level converter HV2 with Arduino TX.
6. Connect Arduino GND with the bi-directional logic level converter GND (high voltage side).

Writing sketches

This section shows you how to write sketches to control the motors with the TB6612FND Arduino library. You can download the library via the following link:

```
https://github.com/sparkfun/SparkFun_TB6612FNG_Arduino_Library/archive/master.zip
```

You can also get the latest version of the library from the GitHub repository:

```
https://github.com/sparkfun/SparkFun_TB6612FNG_Arduino_Library
```

After downloading the library, extract and copy the unzipped folder to the `libraries` folder of your Arduino sketchbook.

Using Blynk

Blynk (`http://www.blynk.cc`) allows you to control Arduino, ESP8266, ESP-12, NodeMCU, Particle Photon, Raspberry Pi, and other microcomputers with the Android or iOS installed smartphone over the internet. It also supports bluetooth and bluetooth low energy connectivity:

1. Download the Blynk app for Android (`https://play.google.com/store/apps/details?id=cc.blynk`) or iOS (`https://itunes.apple.com/us/app/blynk-control-arduino-raspberry/id808760481?ls=1&mt=8`) on your smartphone.
2. After downloading the Blynk app, you'll need to create a new Blynk account by tapping on **Create New Account**. Alternatively, you can log in to the Blynk app with your Facebook account by tapping on **Log In with Facebook** (*Figure 2.20*).

Make sure to provide a real email address or your Blynk account to get the Auth Token by email.

Figure 2.20: Creating a new account for the Blynk app

3. Create a new project by selecting **Create New Project** (*Figure 2.21*):

Figure 2.21: Creating a new project with Blynk

4. Type a name for your project in the **Project Name** box. Then, choose the ARDUINO UNO from the **HARDWARE MODEL** list and tap the **Create** button to create the project (*Figure 2.22*):

Figure 2.22: Project settings

5. After the project has been created, you will get your **Auth Token** via email. Check your email inbox and find the **Auth Token** (*Figure 2.23*):

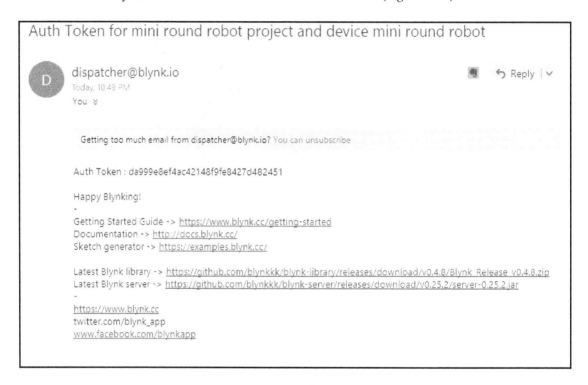

Figure 2.23: Auth Token for the project

6. Add a **Joystick** widget from the **Widget Box** section (*Figure 2.24*):

Figure 2.24: Adding the Joystick widget

7. The joystick widget has been added to the design view (*Figure 2.25*). Note that the x and y coordinates for the center are 128, 128:

Figure 2.25: Joystick widget in the design view

8. Tap on the Joystick widget in the design view to see the **Joystick Settings**.
9. Slide the **SPLIT/MERGE** switch to **Merge** (*Figure 2.26*):

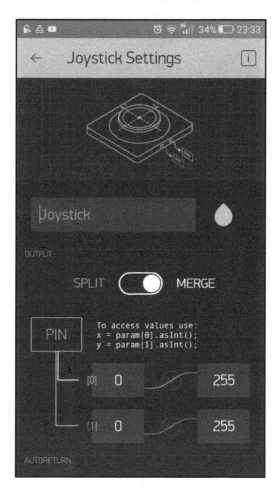

Figure 2.26: Joystick widget settings

10. Set the virtual **PIN** to **V1** and set **AUTORETURN** to **OFF** (*Figure 2.27*):

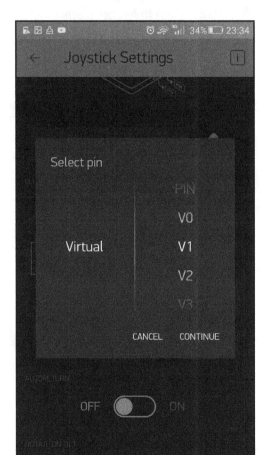

Figure 2.27: Selecting a virtual pin

Installing the Blynk library for Arduino

The following steps will show you how to install the Blynk library on the Arduino IDE:

1. Download the latest release `.zip` file from GitHub (`https://github.com/blynkkk/blynk-library/releases/latest`). You should get a ZIP file named `Blynk_Release_vXX.zip`, where vXX represents the version number.

2. Unzip the archive. You will notice that the archive contains the following folders:
   - `Libraries`
   - `Tools`

3. Inside the `Libraries` folder, you will notice the following folders:
   - `Blynk`
   - `BlynkESP8266_Lib`

4. Copy all these libraries to `your_sketchbook_folder/Libraries` of the Arduino IDE.

 To find the location of `your_sketchbook_folder`, go to the top menu in the Arduino IDE: **File | Preferences** (if you are using MacOS, go to **Arduino | Preferences**).

5. Inside the `Tools` folder, you will notice the following folders:
   - `BlynkUpdater`
   - `BlynkUsbScript`

6. Copy all these tools to `your_sketchbook_folder/tools` of the Arduino IDE.

7. After copying all the necessary libraries and tools, the `sketchbook_folder` should now look as follows:

   your_sketchbook_folder/libraries/Blynk

   your_sketchbook_folder/libraries/BlynkESP8266_Lib

   ...

   your_sketchbook_folder/tools/BlynkUpdater

   your_sketchbook_folder/tools/BlynkUsbScript

   ...

8. Finally, restart the Arduino IDE if already open.

Writing the Arduino sketch

In this section, you will learn how to build the Arduino sketch for the mini round robot, step by step based on the example sketch you generate with the Blynk Sketch Code Builder.

Working with the Blynk Sketch Code Builder

The **Blynk Sketch Code Builder** is an easy-to-use tool for building and writing sketches for targeting specific hardware models. The tool is located at `https://examples.blynk.cc/`.

1. Using your favorite web browser, open the Blynk Sketch Code Builder (*Figure 2.28*):

>

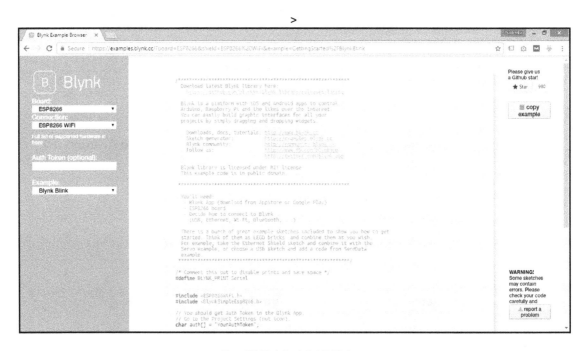

Figure 2.28: Blynk Sketch Code Builder home

2. Choose **Arduino Uno** from the **Board** drop-down list (*Figure 2.29*):

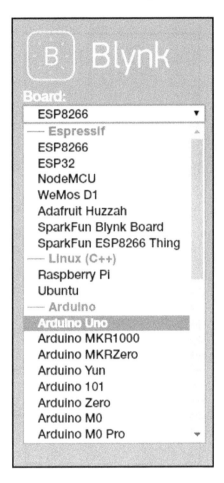

Figure 2.29: Selecting the board type

3. Choose **ESP8266 WiFi Shield** from the **Connection** drop-down list if not already selected (*Figure 2.30*):

Figure 2.30: Selecting the connection type

4. Choose **Virtual Pin Read** from the **Example** drop-down list (*Figure 2.31*):

Figure 2.31: Selecting an example sketch

5. The example sketch will generate and show in the middle section of the window (*Figure 2.32*):

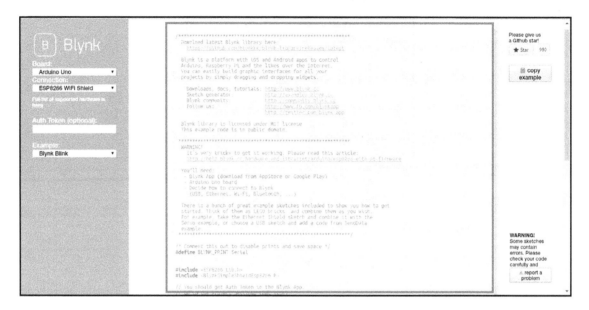

Figure 2.32: Sketch section

6. Copy the code to the clipboard by clicking on the **copy example** button.
7. Paste the code to the Arduino IDE using **Edit | Paste** (menu bar) or *Ctrl + V* (keyboard).

Listing 2-1 – Example code generated with the Blynk Sketch Code Builder

Listing 2-1 shows the example code generated with the Blynk Sketch Code Builder for reading a virtual pin with Arduino UNO and ESP8266. The ESP8266 works as the Wi-Fi shield for the Arduino:

```
/*************************************************************
  Download latest Blynk library here:
    https://github.com/blynkkk/blynk-library/releases/latest

  Blynk is a platform with iOS and Android apps to control
  Arduino, Raspberry Pi and the likes over the Internet.
  You can easily build graphic interfaces for all your
  projects by simply dragging and dropping widgets.

    Downloads, docs, tutorials: http://www.blynk.cc
```

```
   Sketch generator:        http://examples.blynk.cc
   Blynk community:         http://community.blynk.cc
   Follow us:               http://www.fb.com/blynkapp
                            http://twitter.com/blynk_app


 Blynk library is licensed under MIT license
 This example code is in public domain.


 *************************************************************
 WARNING!
    It's very tricky to get it working. Please read this article:
http://help.blynk.cc/hardware-and-libraries/arduino/esp8266-with-at-firmwar
e

 This sketch shows how to read values from Virtual Pins

 App project setup:
    Slider widget (0...100) on Virtual Pin V1
 *************************************************************/

/* Comment this out to disable prints and save space */
#define BLYNK_PRINT Serial


#include <ESP8266_Lib.h>
#include <BlynkSimpleShieldEsp8266.h>

// You should get Auth Token in the Blynk App.
// Go to the Project Settings (nut icon).
char auth[] = "YourAuthToken";

// Your WiFi credentials.
// Set password to "" for open networks.
char ssid[] = "YourNetworkName";
char pass[] = "YourPassword";

// Hardware Serial on Mega, Leonardo, Micro...
#define EspSerial Serial

// or Software Serial on Uno, Nano...
//#include <SoftwareSerial.h>
//SoftwareSerialEspSerial(2, 3); // RX, TX

// Your ESP8266 baud rate:
#define ESP8266_BAUD 115200

ESP8266 wifi(&EspSerial);
```

```
// This function will be called every time Slider Widget
// in Blynk app writes values to the Virtual Pin V1
BLYNK_WRITE(V1)
{
intpinValue = param.asInt(); // assigning incoming value from pin V1 to a
variable

  // process received value
}

void setup()
{
  // Debug console
  Serial.begin(9600);

  // Set ESP8266 baud rate
  EspSerial.begin(ESP8266_BAUD);
  delay(10);

  Blynk.begin(auth, wifi, ssid, pass);
  // You can also specify server:
  //Blynk.begin(auth, wifi, ssid, pass, "blynk-cloud.com", 8442);
  //Blynk.begin(auth, wifi, ssid, pass, IPAddress(192,168,1,100), 8442);
}

void loop()
{
Blynk.run();
}
```

Now, you will modify the preceding example code generated with the Blynk Sketch Code Builder to control the mini round robot. The following steps will describe how to do it:

1. In the example sketch, find this line:

    ```
    char auth[] = "YourAuthToken";
    ```

    Change it with your Auth Token (it should be in your email inbox after you created the project in the Blynk app):

    ```
    char auth[] = "53e4da8793764b6197fc44a673ce4e21";
    ```

2. In the example sketch, find this line:

    ```
    char ssid[] = "YourNetworkName";
    ```

Change it with your Wi-Fi network's SSID:

```
char ssid[] = "Dialog4G";
```

3. In the example sketch, find this line:

```
char pass[] = "YourPassword";
```

Change it with your Wi-Fi network's password:

```
char pass[] = "123456";
```

4. In the example sketch, find this section:

```
BLYNK_WRITE(V1)
{
intpinValue = param.asInt(); // assigning incoming value from
pin V1 to a variable

  // process received value
}
```

Change it to look exactly like this:

```
#include BLYNK_WRITE(V1)
{
intx = param[0].asInt(); // assigning incoming value from pin
V1 to a variable
inty = param[1].asInt(); // assigning incoming value from pin
V1 to a variable

  // process received value

// Do something with x and y
Serial.print("X = ");
Serial.print(x);
Serial.print("; Y = ");
Serial.println(y);
}
```

5. Add the following section just after the previous step:

```
#include <BlynkSimpleShieldEsp8266.h>
// This is the library for the TB6612 that contains the class
Motor and all the functions
#include <SparkFun_TB6612.h>
```

6. Comment the following section:

```
// Hardware Serial on Mega, Leonardo, Micro...
#define EspSerial Serial1
```

Now it should be like as follows:

```
// Hardware Serial on Mega, Leonardo, Micro...
// #define EspSerial Serial1
```

7. Uncomment the following section:

```
// or Software Serial on Uno, Nano...
//#include <SoftwareSerial.h>
//SoftwareSerialEspSerial(2, 3); // RX, TX
```

Now it should be as follows:

```
// or Software Serial on Uno, Nano...
#include <SoftwareSerial.h>
SoftwareSerialEspSerial(2, 3); // RX, TX
```

8. Add the following section just after `ESP8266 wifi(&EspSerial);`:

```
// Pins for all inputs, keep in mind the PWM defines must be on
PWM pins
#define AIN1 12
#define BIN1 7
#define AIN2 4
#define BIN2 8
#define PWMA 5
#define PWMB 6
#define STBY 9

constintoffsetA = 1;
constintoffsetB = 1;

// Initializing motors.
Motor motor1 = Motor(AIN1, AIN2, PWMA, offsetA, STBY);
Motor motor2 = Motor(BIN1, BIN2, PWMB, offsetB, STBY);
```

9. Add the following section to the `BLYNK_WRITE(V1)` function just after the `int y = param[1].asInt();`:

```
if(y>220)
/*
Use of the forward function, which takes as argumentstwo motors
and optionally a speed.  If a negative number is used for speed
it will go backwards
*/
forward(motor1, motor2, 150);

else if(y<35)
/*
Use of the back function, which takes as arguments two motors
and optionally a speed.  Either a positive number or a negative
number for speed will cause it to go backwards
*/
back(motor1, motor2, -150);

else if(x>220)
/*
Use of the left and right functions which take as arguments two
motors and a speed.  This function turns both motors to move in
the appropriate direction.  For turning a single motor use
drive
*/
right(motor1, motor2, 150);

else if(x<35)
/*
Use of the left and right functions which take as arguments two
motors and a speed.  This function turns both motors to move in
the appropriate direction.  For turning a single motor use
drive
*/
left(motor1, motor2, 150);

else
brake(motor1, motor2);
```

Listing 2-2 – complete sketch for mini round robot

Listing 2-2 shows the complete sketch that can be used to control the mini round robot:

```
/****************************************************************
  Download latest Blynk library here:
    https://github.com/blynkkk/blynk-library/releases/latest
  Blynk is a platform with iOS and Android apps to control
  Arduino, Raspberry Pi and the likes over the Internet.
  You can easily build graphic interfaces for all your
  projects by simply dragging and dropping widgets.
    Downloads, docs, tutorials: http://www.blynk.cc
   Sketch generator: http://examples.blynk.cc
   Blynk community: http://community.blynk.cc
   Follow us: http://www.fb.com/blynkapp
   http://twitter.com/blynk_app

  Blynk library is licensed under MIT license
  This example code is in public domain.

  ***************************************************************
  WARNING!
     It's very tricky to get it working. Please read this article:
  http://help.blynk.cc/hardware-and-libraries/arduino/esp8266-with-at-firmwar
  e

  This sketch shows how to read values from Virtual Pins

  App project setup:
    Slider widget (0...100) on Virtual Pin V1
  ***************************************************************/

/* Comment this out to disable prints and save space */
#define BLYNK_PRINT Serial

#include <ESP8266_Lib.h>
#include <BlynkSimpleShieldEsp8266.h>

// This is the library for the TB6612 that contains the class Motor and all
the functions
#include <SparkFun_TB6612.h>

// You should get Auth Token in the Blynk App.
// Go to the Project Settings (nut icon).
char auth[] = "53e4da8793764b6197fc44a673ce4e21";

// Your WiFi credentials.
// Set password to "" for open networks.
```

```
char ssid[] = "Dialog4G";
char pass[] = "123456";

// Hardware Serial on Mega, Leonardo, Micro...
// #define EspSerial Serial

// or Software Serial on Uno, Nano...
#include <SoftwareSerial.h>
SoftwareSerial EspSerial(2, 3); // RX, TX

// Your ESP8266 baud rate:
#define ESP8266_BAUD 115200

ESP8266 wifi(&EspSerial);

// Pins for all inputs, keep in mind the PWM defines must be on PWM pins
#define AIN1 12
#define BIN1 7
#define AIN2 4
#define BIN2 8
#define PWMA 5
#define PWMB 6
#define STBY 9

const int offsetA = 1;
const int offsetB = 1;

// Initializing motors
Motor motor1 = Motor(AIN1, AIN2, PWMA, offsetA, STBY);
Motor motor2 = Motor(BIN1, BIN2, PWMB, offsetB, STBY);

// This function will be called every time Slider Widget
// in Blynk app writes values to the Virtual Pin V1
BLYNK_WRITE(V1)
{
  int x = param[0].asInt(); // get x axis value
  int y = param[1].asInt(); // get y axis value
  // process received value
  // Do something with x and y
  Serial.print("X = ");
  Serial.print(x);
  Serial.print("; Y = ");
  Serial.println(y);

 if (y > 220)
   /*
     Use of the forward function, which takes as arguments two motors
     and optionally a speed. If a negative number is used for speed
```

```
      it will go backwards
   */
   forward(motor1, motor2, 150);

 else if (y < 35)
    /*
      Use of the back function, which takes as arguments two motors
      and optionally a speed. Either a positive number or a negative
      number for speed will cause it to go backwards
    */
    back(motor1, motor2, -150);

 else if (x > 220)
    /*
      Use of the left and right functions which take as arguments two
      motors and a speed. This function turns both motors to move in
      the appropriate direction. For turning a single motor use drive
    */
    right(motor1, motor2, 150);

 else if (x < 35)
    /*
      Use of the left and right functions which take as arguments two
      motors and a speed. This function turns both motors to move in
      the appropriate direction. For turning a single motor use drive
    */
    left(motor1, motor2, 150);

 else
    brake(motor1, motor2);


}
void setup()
{
  // Debug console
  Serial.begin(9600);

  // Set ESP8266 baud rate
  EspSerial.begin(ESP8266_BAUD);
  delay(10);

  Blynk.begin(auth, wifi, ssid, pass);
  // You can also specify server:
  //Blynk.begin(auth, wifi, ssid, pass, "blynk-cloud.com", 8442);
  //Blynk.begin(auth, wifi, ssid, pass, IPAddress(192,168,1,100), 8442);
}
void loop()
```

```
{
  Blynk.run();
}
```

Uploading the sketch

You can upload the sketch from the Arduino IDE to the Arduino UNO board through a USB A-to-B cable. Before uploading the sketch, verify it by clicking on the **Verify** button to see if there are any errors. If you have any errors in your code, debug them and upload the sketch to the Arduino board.

Applying power

Now it's time to power your Mini Round Robot with two battery packs attached to the chassis:

1. Connect the barrel plug of the 9V battery to the DC barrel jack of the Arduino.

2. Then, connect the output from the 4 x AA battery to the motor driver board (*Figure 2.33*):

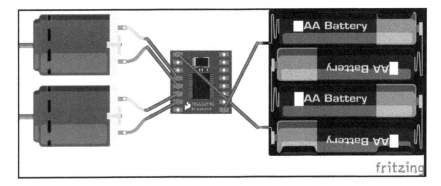

Figure 2.33: Connecting 4 x AA battery to the motor driver board

3. Connect the positive lead of the battery to the VM pin of the motor driver board.
4. Connect the negative lead of the battery to the GND pin of the motor driver board.

Playing the robot

You can test your robot with the project that you have created with the Blynk app:

1. Run the mini round robot project in the Blynk app by tapping on the **Play** button.

2. Move the joystick to forward, backward, left, and right by tapping and holding the center circle (*Figure 2.34*). When you release the center circle, it will automatically come back to the center and apply a break on the mini round robot:

Figure 2.34: The joystick widget in play mode

3. While you move the joystick around, the x and y coordinates will display in the top right corner of the joystick widget.
4. You can fine tune the movements of the robot by changing the x and y coordinate value thresholds mentioned in your Arduino sketch.

Summary

In this chapter, you learned to build a two-wheeled drive mini round robot with a caster ball. While doing this, you understood how to integrate ESP8266 ESP01 and Arduino UNO to control DC motors through a dual H-bridge motor driver. Then, you built a mobile app with Blynk and used a few Arduino libraries to build a complete sketch. Finally, you uploaded the sketch to the Arduino board.

In `Chapter 3`, *Using Encoders*, you will learn how to add wheel encoders to the Mini Round Robot and read values for average speed and distance traveled using a Blynk app.

3
Using Encoders

In this chapter, you will learn how to modify the Mini 3-Layer Round Robot you built in Chapter 2, *Building a Mini Round Robot with Original ESP8266*, by connecting wheel encoders with the motor shaft (not the output shaft) of the DC gear motor to measure the distance traveled and the average speed of the robot chassis.

In this chapter, you will learn the following topics:

- About the SparkFun wheel encoder
- Connecting the wheel encoders to DC gear motors
- Building a Blynk app to display distance traveled and average speed with the Labeled Value widget
- Reading the pulses generated by the two wheel encoders
- Calculating the distance traveled
- Calculating the average speed

Things you will need

You will need the following things to modify the Wi-Fi controlled Mini 3-Layer Round Robot by adding the wheel encoders:

- The Wheel Encoder Kit (https://www.sparkfun.com/products/12629)
- The Blynk app you built in Chapter 2, *Building a Mini Round Robot with Original ESP8266*

The Wheel Encoder Kit

The SparkFun Wheel Encoder Kit (*Figure 3.1*) is a simple add-on to any wheeled robot that can be used to measure the distance traveled and the speed (average speed or instantaneous speed). The Wheel Encoder Kit comes with the following things:

- Two neodymium eight-pole magnets with rubber hubs
- Two hall-effect sensors terminated with 150 mm cables and 3-pin female servo headers:

Figure 3.1: The Wheel Encoder Kit. Image courtesy of SparkFun Electronics (https://www.sparkfun.com)—https://creativecommons.org/licenses/by/2.0/

A neodymium eight-pole magnet has four north poles and four south poles. In other words, it has four pairs of north and south poles. The hall-effect sensor can detect the poles from more than 3 mm (1/8 inch). You can get the number of pulses per wheel revolution with the hall-effect sensor, and it varies depending on the pulse edge or edges you're going to detect (of course you can configure it in your Arduino sketch). It will also depend on the place you will attach the encoder disc to the motor. You can attach the encoder disc to the drive shaft or motor shaft of the DC gear motor (for some DC gear motors, you can only connect the encoder disc to the drive shaft).

As an example, assume that you have a DC gear motor with a 48:1 gear ratio, and you have connected the encoder disc to the motor shaft of the motor. Also, assume that the discs have four north and four south poles. So, for a single rotation of a wheel, it would require 48 rotations of the motor, multiplied by 8 for the poles, that should give you 384 signal changes from the sensor.

A hall-effect sensor has three wires (*Figure 3.2*), typically with the following colors:

- Power – RED
- Ground – BLACK
- Signal – WHITE or any color

A hall-effect sensor requires 3-24V DC with a power supply of 4 mA. Therefore, you can provide the required power to the two hall-effect sensors through the Arduino UNO:

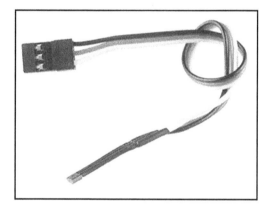

Figure 3.2: The Hall-effect sensor. Image courtesy of SparkFun Electronics (https://www.sparkfun.com)—https://creativecommons.org/licenses/by/2.0/

Connecting the encoders with motors

First, remove the two DC motors from the Mini 3-Layer Round Robot Chassis Kit by disassembling it.

Now you're ready to connect the wheel encoders to the motors. Take the two neodymium magnets and two hall-effect sensors. The neodymium magnet comes with a rubber hub and allows you to press fit it over most small motors and drive shafts used in low-power gearboxes. *Figure 3.3* shows a neodymium magnet connected to the drive shaft of the motor. The hall-effect sensor can be connected to the motor using a 4" zip tie. You can bend the wires to place the hall-effect sensor to face the neodymium magnet. Use about 3 mm of space between the hall-effect sensor and the neodymium magnet.

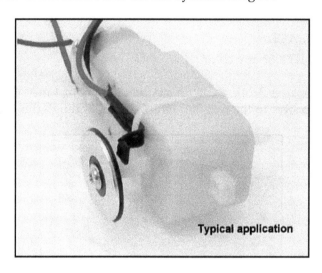

Figure 3.3: Connecting the neodymium magnet to the drive shaft. Image source—https://cdn.sparkfun.com/datasheets/Robotics/multi-chassis%20encoder001.pdf

Wiring the encoders with Arduino

Figure 3.4 shows the wiring diagram for connecting two hall-effect sensors with Arduino UNO:

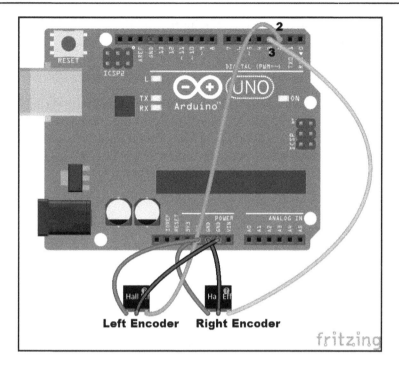

Figure 3.4: Wiring diagram for the encoder

Use the existing circuit you built in `Chapter 2`, *Building a Mini Round Robot with Original ESP8266*, to control the two DC gear motors with SparkFun Dual TB6612FNG (1 A) motor driver, Arduino UNO, ESP8266 (ESP-01), and a Bi-Directional Logic Level Converter. The following steps describe how to connect the two wheel encoders with Arduino UNO:

- Hall-effect sensor attached to the left wheel:
  1. Connect the Signal wire of the hall-effect sensor to Arduino digital pin 2.
  2. Connect the Power wire of the hall-effect sensor to Arduino 5V.
  3. Connect the Ground wire of the hall-effect sensor to Arduino GND.
- Hall-effect sensor attached to the right wheel:
  1. Connect the Signal wire of the hall-effect sensor to Arduino digital pin 3.
  2. Connect the Power wire of the hall-effect sensor to Arduino 5V.
  3. Connect the Ground wire of the hall-effect sensor to Arduino GND.

You can use the Arduino UNO digital pin 2 and 3 as interrupt pins to read pulses coming from the hall-effect sensors.

Great, you have successfully connected the two wheel encoders with Arduino UNO. Now, assemble the 3-Layer Mini Round Robot Chassis to build the robot (use the same instructions provided in Chapter 2, *Building a Mini Round Robot with Original ESP8266*). Make sure that the power switches for both 9V and 4 AA battery boxes are in the **OFF** position before connecting electronic components together.

Reading encoders

A hall-effect sensor varies its output voltage in response to a magnetic field. The eight-pole neodymium magnetic disk attached to the motor shaft can generate a magnetic field around the hall-effect sensor. You can connect the wheel encoders (hall-effect sensor) with Arduino interrupt pins (digital pin 2 and 3 for UNO) to read pulses. You can use either the **RISING**, **FALLING**, or **CHANGE** mode to define when the interruption should be triggered. You will get different results depending on the mode you use.

Creating a Blynk app

The following steps will describe how to modify the Mini Round Robot Blynk app you built in Chapter 2, *Building a Mini Round Robot with Original ESP8266*, to read and display the distance traveled and average speed of the chassis:

1. Open the Mini Round Robot Blynk app (*Figure 3.5*). Switch the app to the **EDIT** mode if it is in the **PLAY** mode:

Figure 3.5: The Mini Round Robot app created in chapter 2, Building a Mini Round Robot with Original ESP8266

2. Tap on the **Widget Box** icon to get the list of widgets. Then, tap on the **Labeled Value** widget under the **DISPLAYS** section (*Figure 3.6*):

Figure 3.6: The Labeled Value widget

3. A **Labeled Value** widget will add to the canvas with a default name, **LABELED**. Now, tap the **Labeled Value** widget to change its settings (*Figure 3.7*):

Figure 3.7: The Labeled Value widget placed on the canvas

4. In the settings page for the **Labeled Value** widget, replace the default widget name **Labeled** with **Distance**. This will provide you a placeholder to display the calculated distance travelled.

5. Under **INPUT,** select the PIN as **Virtual** pin **V2**. Then, tap **CONTINUE** to apply
    the changes (*Figure 3.8*):

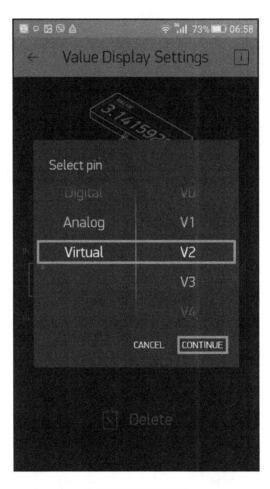

Figure 3.8: Selecting a virtual pin

6. Under **LABEL**, replace the default text, **LABEL** with `Distance: /pin/ in`. The `/pin/` reserves a placeholder for the calculated distance values. Also, `in` indicates the distance in inches. As an instance, the value `50` can be displayed as `Distance: 50 in` (*Figure 3.9*). Keep **READING RATE** as **1 sec**:

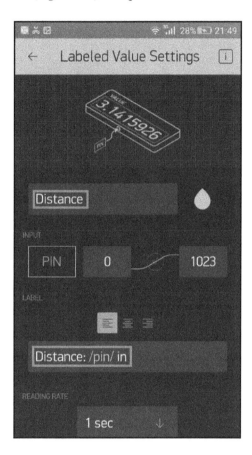

Figure 3.9: The Labeled value settings

7. Tap the back icon on the toolbar to see the changes applied to the **Labeled Value** widget on the canvas (*Figure 3.10*):

Figure 3.10: The configured Labeled Value widget for Distance

8. Repeat the preceding steps to add another **Labeled Value** widget to display **Speed**. Configure it with virtual pin **V3** and change the **LABEL** as Speed: /pin/ in/min (*Figure 3.11*):

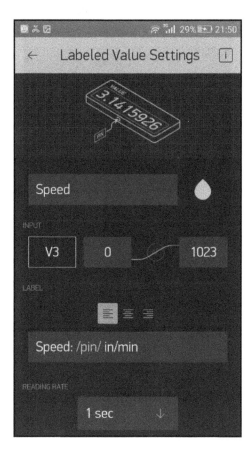

Figure 3.11: Configuring the Labeled Value widget for Speed

Now you have successfully added two **Labeled Value** widgets to the **mini round robot** app (*Figure 3.12*):

Figure 3.12: Modified app to display distance and speed

Writing an Arduino sketch

Listing 3-1 shows an Arduino sketch that you can use to calculate the distance traveled and the average speed based on the pulses coming from the two wheel encoders. Then, it will display the result on the Blynk app by sending data to the **Labeled Value** widgets over the **Virtual Pin** (**V2**) and (**V3**).

Listing 3-1 – Displaying average of the pulses generated by the two wheel encoders (the hall-effect sensors)

The following is the counting of the average pulses generated by the wheel encoder:

```
#include <ESP8266_Lib.h>
#include <BlynkSimpleShieldEsp8266.h>
#include <SparkFun_TB6612.h>

char auth[] = "53e4da8793764b6197fc44a673ce4e21";

char ssid[] = "YourNetworkName";
char pass[] = "YourPassword";

#include <SoftwareSerial.h>
SoftwareSerial EspSerial(2, 3); // RX, TX

#define ESP8266_BAUD 115200

ESP8266 wifi(&EspSerial);

#define AIN1 12
#define BIN1 7
#define AIN2 4
#define BIN2 8
#define PWMA 5
#define PWMB 6
#define STBY 9

const int offsetA = 1;
const int offsetB = 1;

Motor motor1 = Motor(AIN1, AIN2, PWMA, offsetA, STBY);
Motor motor2 = Motor(BIN1, BIN2, PWMB, offsetB, STBY);

/*
   Variable declaration section for encoders
*/
unsigned long time;

volatile int lCount = 0; // ticks counter for left encoder
volatile int rCount = 0; // ticks counter for right encoder
/*

*/

/*
```

```
    Send calculated distance traveled and average speed values to Blynk
Virtual Pins
*/

BlynkTimer timer;

void myTimerEvent()
{
  float averagePulseCount = (lCount + rCount) / 2;

  //Calculate and display distance traveled
  float wheelDiameter = 65;
  float wheelCircumference = wheelDiameter * PI;
  float distanceTraveled = ((wheelCircumference / 8) * averagePulseCount) /
25.4;
  Blynk.virtualWrite(V2, distanceTraveled);


  //Calculate and display average speed
  time = millis(); // Counting time science the robot started
  float minutes = (time / 1000) / 60; //Converts milliseconds to minutes
  float averageSpeed = distanceTraveled / minutes;
  Blynk.virtualWrite(V3, averageSpeed);

}
/*

*/

// This function will be called every time Slider Widget
// in Blynk app writes values to the Virtual Pin V1
BLYNK_WRITE(V1)
{
  int x = param[0].asInt(); // assigning incoming value from pin V1 to a
variable
  int y = param[1].asInt(); // assigning incoming value from pin V1 to a
variable

  if (y > 220)

    forward(motor1, motor2, 150);

  else if (y < 35)

    back(motor1, motor2, -150);

  else if (x > 220)
```

```
    right(motor1, motor2, 150);

  else if (x < 35)

    left(motor1, motor2, 150);

  else
    brake(motor1, motor2);

}

void setup()
{
  // Debug console
  Serial.begin(9600);

  // Set ESP8266 baud rate
  EspSerial.begin(ESP8266_BAUD);
  delay(10);

  Blynk.begin(auth, wifi, ssid, pass);

  /*
    Attach interrupts to Arduino digital pins
  */

  // Setup the myTimerEvent() function to be called every second
  timer.setInterval(1000L, myTimerEvent);

  // Configure digital pin 2 and 3 as input and inverts the behavior of the
  INPUT mode, where HIGH means the sensor is off, and LOW means the sensor is
  on
  pinMode(2, INPUT_PULLUP);
  pinMode(3, INPUT_PULLUP);
  // Attach interrupts
  attachInterrupt(digitalPinToInterrupt(2), leftMotorEncoderCallback,
  CHANGE); // Attach left wheel encoder to the Arduino digital pin 2 for
  trigger a function using interrupts
  attachInterrupt(digitalPinToInterrupt(3), rightMotorEncoderCallback,
  CHANGE); // Attach right wheel encoder to the Arduino digital pin 3 for
  trigger a function using interrupts

  /*

  */
}
```

```
void loop()
{
  Blynk.run();
}
/*
   Callback functions for counting signal (pulses) generated by the left
and right wheel encoders
*/

void leftMotorEncoderCallback()
{
  lCount++;
}

void rightMotorEncoderCallback()
{
  rCount++;
}
/*

*/
```

Let's examine the preceding Arduino sketch and learn how to trigger a user defined function to count the pulses generated by a wheel encoder (hall-effect sensor) and send the calculated values for distance traveled to the Blynk app over a **Virtual Pin**:

- First, declare the following two variables to store pulses for the left and right wheel encoders:

    ```
    volatile int lCount = 0; // Pulse counter for left encoder
    volatile int rCount = 0; // Pulse counter for right encoder
    ```

- The `leftMotorEncoderCallback()` function returns the number of pulses generated by the left wheel encoder. The `lcount` variable will increase each time you call the `leftMotorEncoderCallback()` function with the `pinMode()` function that will be discussed in the next section. Also, the `rightMotorEncoderCallback()` function returns the number of pulses generated by the right wheel encoder:

    ```
    /*
       Callback functions for counting signal (pulses) generated by
    the left and right wheel encoders
       */

    void leftMotorEncoderCallback()
    {
    ```

```
      lCount++;
    }

    void rightMotorEncoderCallback()
    {
      rCount++;
    }
    /*
    *
    */
```

- The wheel encoders are attached to the Arduino digital pin 2 and 3 with the INPUT_PULLUP parameter using the pinMode() function. The two functions, leftMotorEncoderCallback() and rightMotorEncoderCallback(), are triggered by digitals pin 2 and 3 with the attachInterrupt() function. The Arduino sketch only records a signal from the sensor when it detects a **CHANGE** in signal (detects both **RISING** and **FALLING**). The constant CHANGE is used to trigger the interrupt whenever the pin changes its value. The setInterval() function is used to set up the myTimerEvent() function to be called every second. It accepts the time in milliseconds. The L indicates that the variable type is integer long:

```
    /*
       Attach interrupts to Arduino digital pins
       */
       // Setup the myTimerEvent() function to be called every
    second
       timer.setInterval(1000L, myTimerEvent);

       // Configure digital pin 2 and 3 as input and inverts the
    behavior of the INPUT mode, where HIGH means the sensor is off,
    and LOW means the sensor is on
       pinMode(2, INPUT_PULLUP);
       pinMode(3, INPUT_PULLUP);
       // Attach interrupts
       attachInterrupt(digitalPinToInterrupt(2),
    leftMotorEncoderCallback(), CHANGE);   // Attach left wheel
    encoder to the Arduino digital pin 2 for trigger a function
    using interrupts
       attachInterrupt(digitalPinToInterrupt(3),
    rightMotorEncoderCallback(), CHANGE);   // Attach right wheel
    encoder to the Arduino digital pin 3 for trigger a function
    using interrupts
    /*
    *
    */
```

- The `myTimerEvent()` function sends the calculated distance traveled and average speed values every second to the Blynk virtual pin (2) and (3). The `averagePulseCount = (lCount + rCount)/2` formula can be used to calculate the average number of pulses. The result is stored in the variable `averagePulseCount` as a float value:

```
/*
 Send calculated distance traveled and average speed values to
Blynk Virtual Pins
 */
BlynkTimer timer;
void myTimerEvent()
{
   float averagePulseCount=(lCount + rCount)/2;
}
/*
 *
 */
```

Now that your Blynk app can display the average pulse count in the **Labeled Value** widget, distance. In the next section, you will learn how to calculate the distance traveled using the average pulse count.

Calculating the distance traveled

The average pulse count can be used to calculate the approximate distance the chassis travels. Before calculating the distance traveled, you should have the diameter or radius of the wheel attached to the output shaft of the DC gear motor. First, calculate the circumference of the wheel by multiplying the wheel diameter (or radius x 2) with the constant *PI (22/7)*:

Wheel circumference = Wheel diameter x PI

Alternatively, we can use the following formula:

Wheel circumference = 2 x PI x Wheel radius

where *PI = 22/7*

The magnetic disc should be connected to the drive shaft of the motor. A magnetic disc has four north and four south poles. So, for a single rotation of the wheel, it will give eight signal changes from the sensor. The **CHANGE** parameter in the `attachInterrupt()` function detects both **RISING** and **FALLING** edge signal. In other words, it detects all the eight poles in the magnetic disc.

Finally, you can calculate the distance as follows:

*Distance traveled = (wheel circumference / pulses per revolution) * average pulse count*

As an example, assume that you have the following result with the preceding Arduino sketch:

Average pulse count: 1000

Wheel diameter = 65 mm

The distance, d, can be calculated as follows:

Each encoder outputs eight pulses for a wheel revolution; therefore, the robot drives 65 mm. Now, you can calculate the distance for 1000 pulses, as shown in the following formula:

*Distance traveled = (wheel circumference / pulses per revolution) * average pulse count*
               *= (65 / 8) * 1000*
               = 8125 mm
               = 8125 / 25.4
               = 319.88 inches

The distance traveled can be calculated inside the `myTimerEvent()` function with the preceding formula:

You can easily convert millimeters to inches by dividing with 25.4.

```
void myTimerEvent()
{
   float averagePulseCount=(lCount + rCount)/2;
      //Calculate and display distance traveled
   float wheelDiameter = 65;
   float wheelCircumference = wheelDiameter * PI;
   float distanceTraveled = ((wheelCircumference / 8) * averagePulseCount) /
25.4;

   Blynk.virtualWrite(V2, distanceTraveled);
}
```

The `virtualWrite` function will write the result stored in the
`distanceTraveled` variable to the Blynk virtual pin V2 and display on the **Labeled Value**
widget, distance.

Calculating the average speed

The average speed of the chassis can be easily calculated with the distance traveled and the
elapsed time since start. The following formula can be used to calculate the average speed
(the symbol *v* is used for speed, *s* is used for distance traveled, and *t* is used for elapsed
time):

$$v = \frac{s}{t}$$

Let's assume that the Mini Round Robot chassis took *4 minutes* to travel the distance of *2000
inches*. The average speed can be calculated with the preceding formula as follows:

Average Speed = 2000 inches / 4 minutes

    =500 inches per minute

Alternatively, it can be calculated as follows:

Average Speed = 2000 inches / 240 seconds

    = 8.33 inches per second

You can use the labeled value widget that you created to display the average speed on the
Blynk app. First, you will write the calculated average speed to the Blynk virtual pin (v3)
and then, send it back to the labeled value widget, speed.

Listing 3-3 – Calculating and displaying the average speed

The sketch snippet for calculating and displaying the speed is as follows:

```
void myTimerEvent()
{
...
//Calculate and display average speed
   time = millis(); // Counting time science the robot started
float minutes = (time / 1000)/60; //Converts milliseconds to minutes
   float averageSpeed = distanceTraveled / minutes;
   Blynk.virtualWrite(V3, averageSpeed);
}
```

Also, remember to add the following line to the variable declaration section of the sketch:

```
unsigned long time;
```

The `millis()` function returns the number of milliseconds since the program started. You should convert the milliseconds to minutes before calculating the speed. The Blynk virtual pin `V3` is used to write the average speed to the labeled value widget, speed as inches per minute.

Playing the 3-Layer Mini Round Robot

Now, you're ready to play with your 3-Layer Mini Round Robot with the Blynk app that you have developed.

First, slide the power switches in both 9V and 4 AA battery boxes to the **ON** postion. Then, open your Blynk app and tap the **PLAY** button on the toolbar. After establishing the connection between the Blynk app and the robot over Wi-Fi, tap the **Joystick** widget to move the robot. While moving the robot, you can see the distance traveled and average speed values updating on the Blynk app every 1 second.

Summary

In this chapter, you learned how to connect SparkFun Wheel Encoder Kit to the DC gear motors to read and count pulses generated in the hall-effect sensors. Then, you modified the Mini Round Robot Blynk app by adding the labeled value widget to display the distance traveled and average speed of the robot chassis. After this, you calculated the distance traveled and the average speed of the robot chassis based on the pulses generated by the two magnetic discs of the wheel encoders. Finally, you fed the resulting values to the Blynk app through the Blynk virtual pins every 1000 milliseconds.

Further, you can improve the app by calculating and displaying the instantaneous speed rather than displaying the average speed of the chassis.

However, you can build a more compact and lightweight robot by replacing both Arduino UNO and ESP8266 (ESP01) with Adafruit Feather HUZZAH ESP8266. In Chapter 4, *Building a Mini Round Robot with the Feather HUZZAH ESP8266*, you will use Adafruit Feather HUZZAH ESP8266 to rebuild the 3-Layer Mini Round Robot you built in Chapter 2, *Building a Mini Round Robot with Original ESP8266*.

4

Building a Mini Round Robot with the Feather HUZZAH ESP8266

In the previous two chapters, you used the original ESP8266, which is ESP01 and the Arduino UNO to drive motors with a motor driver. You should have noticed that the ESP01 has only two **general-purpose input/output** (**GPIO**) pins. Using two GPIO pins, you can only drive a single motor through a motor driver. Therefore, for two motors, you will need four GPIO pins. As a solution, you used the Arduino UNO as a bridge between the ESP01 and motor driver. In the previous two projects, the ESP01 was used to provide all networking facilities through Wi-Fi and the Arduino UNO worked as the microcontroller.

In this chapter, you will learn the following topics:

- About the Feather HUZZAH ESP8266
- About the DC Motor + Stepper FeatherWing
- Assembling the Mini Round Robot Chassis Kit with Feather HUZZAH ESP8266 and DC Motor + Stepper FeatherWing
- Creating a Blynk app with a Joystick widget
- Writing an Arduino sketch to communicate with the Blynk app and control the robot
- Testing the robot

Things you will need

The following list shows the things that you will need to build the robot. In `Chapter 1`, *Getting Ready* and `Chapter 2`, *Building a Mini Round Robot with Original ESP8266*, you assembled the Mini 3-Layer Round Robot Chassis Kit with the Arduino UNO, a motor driver, ESP8266 ESP01, wheel encoders, and a bi-directional logic level converter.

You can use the same chassis kit by removing all the things assembled on the top chassis layer (the Arduino UNO, motor driver, ESP8266 ESP01, and the bi-directional logic level converter) and the 9V battery holder attached to the bottom chassis layer.

If you want to build a robot with another Mini 3-Layer Round Robot Chassis Kit, first assemble it on the top chassis layer without attaching a 9V battery holder onto the bottom chassis layer:

- 1 x Mini 3-Layer Round Robot Chassis Kit—2WD with DC Motors (`https://www.adafruit.com/product/3244`)
- 1 x Assembled Adafruit Feather HUZZAH with ESP8266 Wi-Fi with Headers (`https://www.adafruit.com/product/3046`)
- 1 x DC Motor + Stepper FeatherWing Add-on For All Feather Boards (https://www.adafruit.com/product/2927)
- 1 x Feather Stacking Headers: 12-pin and 16-pin Female Headers (`https://www.adafruit.com/product/2830`)
- 1 x FeatherWing Doubler - Prototyping Add-on For All Feather Boards (`https://www.adafruit.com/product/2890`)
- 1 x Feather Header Kit - 12-pin and 16-pin Female Header Set (`https://www.adafruit.com/product/2886`)
- 1 x Lithium Ion Battery - 3.7v 2000mAh (`https://www.adafruit.com/product/2011`)
- 1 x 4 x AA Battery Holder with On/Off Switch (`https://www.adafruit.com/product/830`)
- 1 x Button Hex Machine Screw - M4 thread - 8mm long -(one pack includes 50) (`https://www.adafruit.com/product/1160`)
- 1 x Brass M2.5 Standoffs for Pi HATs - Black Plated (One Pack Includes 2) (`https://www.adafruit.com/product/2336`)

- 1 x Alkaline AA batteries (LR6) - 4 pack (https://www.adafruit.com/product/3349)
- 1 x Adafruit Pocket Screwdriver - Black (https://www.adafruit.com/product/3284)
- 1 x Piece of Double-Stick Foam Tape

What is Feather HUZZAH ESP8266?

The Feather HUZZAH ESP8266 is a lightweight breakout board that can be used to build projects without using an additional microcontroller board. It has an onboard USB-serial converter that allows you to upload programs through a USB cable to a computer. The built-in LiPoly battery charger allows you to charge the battery with USB power. The onboard **Reset** button is used to reset the microcontroller.

The Feather HUZZAH ESP8266 exposes the following pins through the breakout board (*Figure 4.1*):

- GPIO pins (0, 2, 4, 5, 12, 13, 14, 15, 16)
- Serial pins (TX, RX)
- I2C pins (SDA, SCL)
- ISP pins (SCK, MOSI, MISO)
- Power pins (GND, BAT, USB, EN, 3V)
- Analog pins (ADC)
- Control pins (RST, CHPD)
- Not-Connected pins (marked with NC)

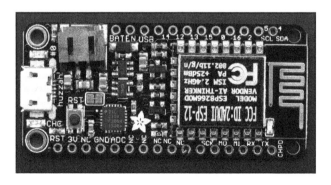

Figure 4.1: Pinout of the Feather HUZZAH ESP8266. Image courtesy of Adafruit Industries—https://www.adafruit.com/

The DC Motor + Stepper FeatherWing

The **DC Motor + Stepper FeatherWing** allows you to drive the two DC motors that are attached to the bottom chassis layer. *Figure 4.2* shows the pinout of the DC Motor + Stepper FeatherWing, which includes the following things:

- **Motor power pins**: Supply 5-12V DC through these pins to drive the motors
- **Motor outputs**: You can connect up to 4 DC gear motors with M1, M2, M3, and M4 terminal blocks
- **I2C pins**: SDA and SCL

Figure 4.2: Pinout of the DC Motor + Stepper FeatherWing. Image courtesy of Adafruit Industries—https://www.adafruit.com

Downloading the Adafruit Motor Shield V2 library

First, you should download the DC Motor library from the Adafruit Motor Shield V2 library at
`https://github.com/ladyada/Adafruit_Motor_Shield_V2_Library/archive/master.zip`.
Then, copy the folder inside the ZIP file to the libraries folder inside your `Arduino UNO Sketchbook` folder and rename it `Adafruit_Motorshield`. After renaming the folder, restart the Arduino UNO IDE for the `Adafruit_Motorshield` library to take effect.

Assembling the robot

You will need the Mini 3-Layer Round Robot Chassis Kit assembled up to the top chassis layer, as discussed in the assembly instructions provided in `Chapter 1`, *Getting Ready*.

The following steps will guide you through assembling the other parts and wiring to build the complete robot.

1. Take the **FeatherWing Doubler** (*Figure 4.3*). The FeatherWing Doubler PCB comes with the following things:
   - One set of Feather Stacking Headers
   - One set of Feather Female Headers

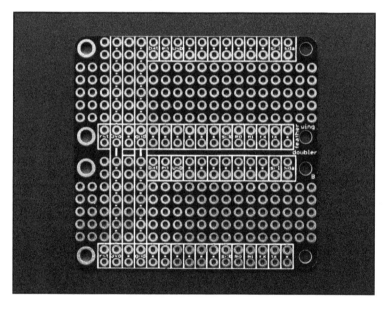

Figure 4.3: The FeatherWing Doubler PCB. Image courtesy of Adafruit Industries—https://www.adafruit.com

2. Solder the **Feather Female Headers** to the FeatherWing Doubler PCB. *Figure 4.4* shows the FeatherWing Doubler PCB with soldered Feather Female Headers:

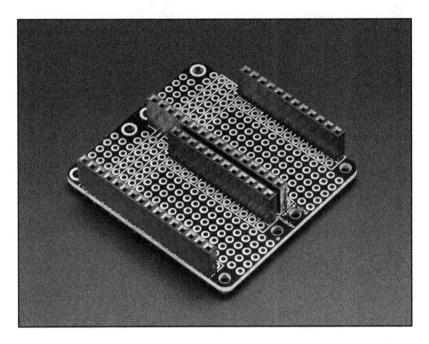

Figure 4.4: The FeatherWing Doubler PCB with soldered Feather Female Headers. Image courtesy of Adafruit Industries—https://www.adafruit.com

3. The FeatherWing Doubler stacks and internally connects the two Feather boards horizontally. This is same as stacking one Feather board on top of another Feather board like shield stacking on the Arduino UNO or hat stacking on the Raspberry Pi.

4. Using four brass standoffs, attach the Doubler PCB to the top chassis layer and secure it with small screws.

5. Stack the Feather HUZZAH and the DC Motor + Stepper FeatherWing on top of the FeatherDoubler (*Figure 4.5*):

Figure 4.5: Stacking the Feather HUZZAH and Motor shield on top of the FeatherDoubler. Image courtesy of Adafruit Industries—https://www.adafruit.com

The two Feather Boards communicate through the I2C bus and control the two motors. The SDA and SCL pins of the two Feather Boards are internally connected through the FeatherDoubler. The DC Motor + Stepper FeatherWing takes 3.3V power from the Feather HUZZAH.

6. Cut a small strip of double-sided tape and place it on the side of the LiPoly battery. Place the battery close to the Feather HUZZAH ESP8266 and stick it to the chassis.

7. The 3.7V LiPoly battery comes with a two-pin JST-PH connector. Connect it to the battery connector of the Feather HUZZAH. The 3.7V LiPoly battery provides power for both the Feather HUZZAH and the DC Motor + Stepper FeatherWing.

8. Connect the left motor to the M4 connector of the motor driver. Use the small flat-head screwdriver to connect the leads from the motor with the connector.

9. Connect the right motor to the M3 connector of the motor driver.

10. Connect the power leads of the 4 x AA battery pack to the power connector of the motor driver. Make sure to turn off the switch before connecting with the motor driver. The battery pack provides about 6V from fresh alkaline batteries. Each battery provides 1.5V and the capacity is 3000 mAh.

11. The fully assembled robot with all the required parts is ready. Still, your robot doesn't have a brain!

Creating a Blynk app

You will need a Blynk app to control the robot with a Joystick widget. You can use the Blynk app you built in `Chapter 1`, *Getting Ready* and use the same Auth Token for the Arduino UNO sketch.

However, first make sure that the following things are correctly configured in your app:

- The **OUTPUT** of the joystick should be set to the **MERGE** mode (*Figure 4.6*). In the **MERGE** mode, the joystick sends one message, consisting of an array of values, to the microcontroller hardware so that you can parse the values on the hardware:

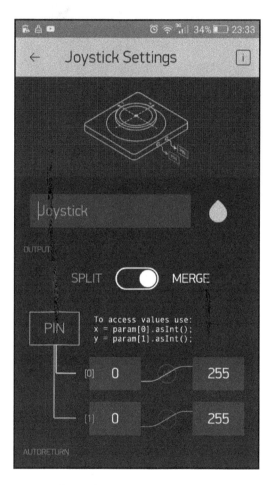

Figure 4.6: Setting the OUTPUT to MERGE mode

- **MERGE** mode can be used with virtual pins only. Use the virtual pin **V1**. This will allow the joystick widget to write to pin **V1** (*Figure 4.7*):

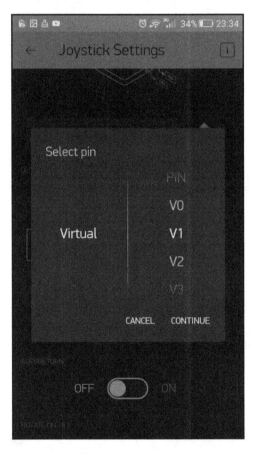

Figure 4.7: Selecting the virtual pin V1

- **ROTATE ON TILT** should be **ON** for the joystick to automatically rotate if you use your smartphone in landscape orientation (*Figure 4.8*).
- **AUTORETURN** should be **OFF**. When it is **ON**, the joystick handle will return to its center position after you release it. When it is **OFF**, it will stay where you left it (*Figure 4.8*):

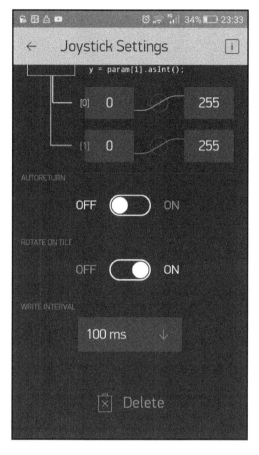

Figure 4.8: Setting AUTORETURN and ROTATE ON TILT

Writing the Arduino UNO sketch

First, generate the Arduino UNO sketch that shows how to read values from virtual pins with the Adafruit HUZZAH and ESP8266 Wi-Fi by using the Blynk Sketch Builder (`https:/ /examples.blynk.cc`). The following instructions will show you how to generate the example sketch with the Blynk Sketch Builder (*Figure 4.9*):

1. Select **Adafruit Huzzah** from the **Board** drop down list.
2. Select **ESP8266 WiFi** from the **Connection** drop down list.
3. Type the **Auth Token** associated with your Blynk app in the **Auth Token (optional)** text box.

4. Select **Virtual Pin Read** from the **Example** drop down list:

Figure 4.9: Configuration steps to generate the sketch for a Virtual Pin Read

5. Click the **Copy example** button in the Blynk Sketch Builder to copy the code onto the clipboard.
6. Then, start the Arduino UNO IDE, create a new file, delete the default sketch, and paste the example code.

The example code generated with the Blynk Sketch Builder can be modified to control the two motors with the Joystick widget through the Adafruit Stepper + DC Motor FeatherWing.

Listing 4-1 – Arduino sketch for controlling the Mini Round Robot

Listing 4-1 presents the complete Arduino sketch that can be used to control the Mini Round Robot with the Joystick widget on the Blynk app:

```
#include <ESP8266WiFi.h>
#include <BlynkSimpleEsp8266.h>

#include <Wire.h>
#include <Adafruit_MotorShield.h>

// You should get Auth Token in the Blynk App.
// Go to the Project Settings (nut icon).
char auth[] = "da999e8ef4ac42148f9fe8427dxxxxxx";

// Your WiFi credentials.
// Set password to "" for open networks.
char ssid[] = "xxxxxx";
char pass[] = "xxxxxx";

Adafruit_MotorShield AFMS = Adafruit_MotorShield();

Adafruit_DCMotor *L_MOTOR = AFMS.getMotor(4);
Adafruit_DCMotor *R_MOTOR = AFMS.getMotor(3);

// This function will be called every time Joystick Widget
// in Blynk app writes values to the Virtual Pin V1
BLYNK_WRITE(V1)
{
  int x = param[0].asInt(); // get x axis value
  int y = param[1].asInt(); // get y axis value

  // process received value
  if (y > 220)
    forward();

  else if (y < 35)
    backward();

  else if (x > 220)
    right();

  else if (x < 35)
    left();

  else
```

```
      stop();
}

void setup()
{
  Blynk.begin(auth, ssid, pass);

  AFMS.begin();

}

void loop()
{
  Blynk.run();
}

void forward() {

  L_MOTOR->setSpeed(200);
  L_MOTOR->run( FORWARD );
  R_MOTOR->setSpeed(200);
  R_MOTOR->run( FORWARD );

}

void backward() {

  L_MOTOR->setSpeed(150);
  L_MOTOR->run( BACKWARD );
  R_MOTOR->setSpeed(150);
  R_MOTOR->run( BACKWARD );

}

void stop() {

  L_MOTOR->setSpeed(0);
  L_MOTOR->run( RELEASE );
  R_MOTOR->setSpeed(0);
  R_MOTOR->run( RELEASE );

}

void left() {

  L_MOTOR->setSpeed(100);
  L_MOTOR->run( BACKWARD );
  R_MOTOR->setSpeed(100);
```

```
   R_MOTOR->run ( FORWARD );

}

void right () {

  L_MOTOR->setSpeed(100);
  L_MOTOR->run ( FORWARD );
  R_MOTOR->setSpeed(100);
  R_MOTOR->run ( BACKWARD );

}
```

The following steps will explain you about the most important configuration settings, modifications, library functions, and user defined functions in the preceding code:

1. Include all the required libraries for Adafruit Stepper + DC Motor FeatherWing and the Feather HUZZAH:

   ```
   #include <Wire.h>
   #include <Adafruit_MotorShield.h>
   ```

   The `Wire.h` library allows you to communicate between Adafruit Stepper + DC Motor FeatherWing and the Feather HUZZAH with I2C. The `Adafruit_MotorShield.h` library provides all the required functions to work with DC motors.

2. Create an instance of `Adafruit_MotorShield` with the default I2C address. The default I2C address is `0x60`:

   ```
   Adafruit_MotorShield AFMS = Adafruit_MotorShield();
   ```

3. Create instances for the two DC motors and connect them to ports M3 and M4.

```
Adafruit_DCMotor *L_MOTOR = AFMS.getMotor(4);
Adafruit_DCMotor *R_MOTOR = AFMS.getMotor(3);
```

4. Inside the `setup()` function of the sketch, initialize the `Adafruit_MotorShield` library:

```
AFMS.begin();
```

5. Add the `forward()` function to move the robot chassis forward:

```
void forward()
{

    L_MOTOR->setSpeed(200);
    L_MOTOR->run( FORWARD );

    R_MOTOR->setSpeed(200);
    R_MOTOR->run( FORWARD );

}
```

6. First, set the `speed` of each motor to `200` using `setSpeed(speed)`, where the `speed` ranges from 0 (stopped) to 255 (full speed). The speed of the motor is controlled by the **pulse width modulation (PWM)** setting. Then, run each motor using `run(direction)` where `direction` is FORWARD.

7. Add `backward()` to move the robot chassis backward:

```
void backward()
{

    L_MOTOR->setSpeed(150);
    L_MOTOR->run( BACKWARD );

    R_MOTOR->setSpeed(150);
    R_MOTOR->run( BACKWARD );

}
```

Like the `forward()` function, first set the `speed` of each motor to `150` using `setSpeed(speed)`. Then, run each motor using `run(direction)`, where `direction` is `BACKWARD`.

8. Add the `stop()` function to turn off both motors and apply the brakes:

```
void stop()
{

  L_MOTOR->setSpeed(0);
  L_MOTOR->run( RELEASE );

  R_MOTOR->setSpeed(0);
  R_MOTOR->run( RELEASE );

}
```

First, set the `speed` of each motor to `0` using `setSpeed(speed)`. This will stop the rotation of the motor's drive shaft. Then turn off each motor using `run(direction)`, where `direction` is `RELEASE`.

9. Add the `left()` function to turn the robot chassis left:

```
void left()
{

  L_MOTOR->setSpeed(100);
  L_MOTOR->run( BACKWARD );

  R_MOTOR->setSpeed(100);
  R_MOTOR->run( FORWARD );

}
```

First, set the `speed` of each motor to `100` using `setSpeed(speed)`. Then, set the left motor to `BACKWARD` and the right motor to `FORWARD` using the `run(direction)` function.

This will create a POINT TURN to the left where the center of rotation is centered between the driving wheels.

For the POINT TURN, one wheel must go forward while the other wheel goes backward. The chassis will turn to the side of the wheel going backward.

Figure 4.10 shows an example of the POINT TURN:

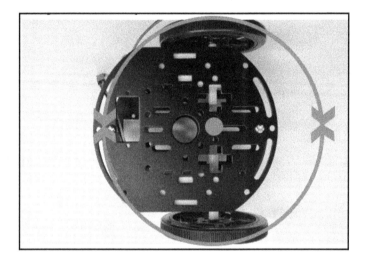

Figure 4.10: Point turn

10. Add the right() function to turn the robot chassis to the right:

```
void right()
{


    L_MOTOR->setSpeed(100);
    L_MOTOR->run( FORWARD );


    R_MOTOR->setSpeed(100);
    R_MOTOR->run( BACKWARD );


}
```

First, set the `speed` of each motor to `100` using `setSpeed(speed)`. Then, set the right motor to `BACKWARD` and the left motor to `FORWARD` using the `run(direction)` function. This will create a `POINT TURN` to the right.

11. Modify the `BLYNK_WRITE(V1)` as shown here to get the value sent by the joystick. You have configured the joystick to send x and y values using the `MERGE` mode to the pin on the ESP8266. When the `MERGE` mode is selected, you are sending just one message, consisting of an array of values. But you'll need to parse it on the hardware. `param[0]` and `param[1]` hold the x and y values sent by the joystick. Remember, the `MERGE` mode can be used with virtual pins only:

```
BLYNK_WRITE(V1)
{
    int x = param[0].asInt(); // get x axis value
    int y = param[1].asInt(); // get y axis value
}
```

12. Then, call each function to move the robot according to the movements in the four directions of the joystick, as follows:

```
if(y>220)
forward();


else if(y<35)
back();


else if(x>220)
right();


else if(x<35)
left();


else
stop ();
```

13. Replace `YourAuthToken` with the Auth Token associated with your Blynk app:

```
// You should get Auth Token in the Blynk App.
// Go to the Project Settings (nut icon).
char auth[] = "YourAuthToken";
```

14. Type the network name and password of your Wi-Fi network. You can use a blank password for open networks:

```
// Your WiFi credentials.
// Set password to "" for open networks.
char ssid[] = "YourNetworkName";
char pass[] = "YourPassword";
```

After modifying the sketch, upload it in to the Feather HUZZAH ESP8266 board using the Arduino IDE. Before uploading or verifying the sketch, remember to choose **Tools | Board** and then, Adafruit HUZZAH ESP8266 on the toolbar.

Testing the robot

This test allows you to fine-tune and correct the rotating directions of the motors:

1. Slide the power switch of the 4 x AA battery case to supply the power to the motors through the motor driver.
2. Open your Blynk app and allow it to connect with the robot.
3. Initially, the joystick handle will be at the center position. Move the handle forward and hold to drive the robot forward. Sometimes, the motors may not rotate in the forward direction and result in an erratic behavior. You can change the direction by simply swapping the two wires from the motor to the motor shield. Once you fix the forward direction of the motors, the other directions will also be fixed.
4. Move the handle to the left. The robot will make a point turn to the left continuously until you release or move the handle to a different position.
5. Release the handle to automatically return it to the center position. The robot will stop immediately.
6. Move the joystick handle backward and hold to drive the robot backward. Then, make a point turn to the right.
7. If you want to change the speed for either direction, you can change the PWM values in the sketch.

Summary

In this chapter, you built the hardware assembly for the Mini Round Robot with the Adafruit Feather HUZZAH ESP8266 and the DC Motor + Stepper FeatherWing. You also built a mobile app with Blynk and wrote the Arduino UNO sketch to control the robot through a Wi-Fi network.

As a simple improvement, you can add a slider widget to the Blynk app to control the speed of your robot. See how it turns to the left or right at its maximum speed!

In Chapter 5, *Line-Following Zumo Robot*, you will build a line following robot using the Zumo Chassis Kit with Adafruit Feather HUZZAH ESP8266.

5
Line-Following Zumo Robot

By now, you know how to connect a robot to a control application through a Wi-Fi network using ESP8266. In this chapter, you will go further to build a robot that follows a line, either a black line on a white surface or a white line on a black surface. Line-following robots usually use an array of IR (Infrared) sensors to detect the line.

In this chapter, you will learn the following topics:

- Assembling the Zumo Chassis Kit
- Connecting SparkFun Line Follower Array with Zumo chassis
- Connecting electronics to detect the line
- Building a line-following course
- Writing an Arduino sketch to control motors according to sensor values

Things you will need

First, let's prepare all the components that you will need to build the line-following robot:

- 1 x Zumo Chasis Kit (no motors) (`https://www.pololu.com/product/1418`)
- 2 x 100:1 Micro Metal Gearmotor HP 6V (`https://www.pololu.com/product/2191`)
- 1 x Basic Sumo Blade for Zumo chassis (optional) (`https://www.pololu.com/product/1410`)
- 4 x Rechargeable NiMH AA battery: 1.2 V, 2200 mAh, 1 cell (`https://www.pololu.com/product/1003`)
- 1 x FeatherWing Doubler - Prototyping Add-on For All Feather Boards (`https://www.adafruit.com/product/2890`)

- 1 x Assembled Feather HUZZAH w/ ESP8266 Wi-Fi with Stacking Headers (https://www.adafruit.com/product/3213)
- 1 x DC Motor + Stepper FeatherWing Add-on For All Feather Boards (https://www.adafruit.com/product/2927)
- 1 x Lithium Ion Battery - 3.7V 2000mAh (https://www.adafruit.com/product/2011)
- 1 x Extra-long break-away 0.1" 16-pin strip male header (5 pieces) (https://www.adafruit.com/product/400)
- 1 x Ceramic capacitor 3-pack 0.1uF 100V (https://www.pololu.com/product/1166)
- 1 x Little Rubber Bumper Feet - Pack of 4 (https://www.adafruit.com/product/550)

Assembling the Zumo chassis Kit

Pololu's Zumo chassis kit contains the components you will need to build a small, high-performance tracked robot platform. Tracked robots are a type of robot that use treads or caterpillar tracks instead of wheels. The kit includes the following components:

1. The Zumo chassis main body
2. 1/16" black acrylic mounting plate
3. Two drive sprockets
4. Two idler sprockets
5. Two 22-tooth silicone tracks
6. Two shoulder bolts with washers and M3 nuts
7. Four 1/4" #2-56 screws and nuts
8. Battery terminals

Figure 5.1 shows the components of the Zumo chassis kit:

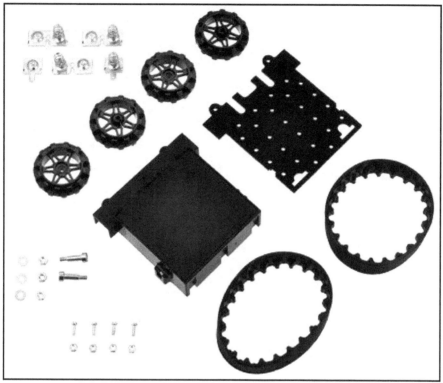

Figure 5.1: Components of the Zumo chassis kit. Image courtesy of Pololu (https://www.pololu.com/)

Additionally, you will need the following things that are not included in the Zumo chassis kit:

- Two 75:1 HP micro metal gearmotors
- Four rechargeable NiMH AA batteries

Normally, you can convert the Zumo chassis to a line-following robot with the following things listed recommended by Pololu (`https://www.pololu.com/`):

- Zumo Shield for Arduino, v1.2 (`https://www.pololu.com/product/2508`)
- Arduino UNO R3 (`https://www.pololu.com/product/2191`)
- Zumo Reflectance Sensor Array (`https://www.pololu.com/product/1419`)

However, in this chapter, you will build a Zumo line-following robot based on the **Adafruit FEATHER HUZZAH ESP8266** and the **DC Motor + Stepper FeatherWing Add-on** board. This allows you to replace the Zumo Shield for the Arduino and Arduino UNO to reduce the vertical space and weight. You will also use the **SparkFun Line Follower Array** (https://www.sparkfun.com/products/13582) instead of the **Zumo Reflectance Sensor Array** (https://www.pololu.com/product/1419) from Pololu.

The assembly instruction for the Zumo chassis kit can be found at https://www.pololu.com/docs/0J54/all. Follow the instructions given in the assembly guide to build the chassis by assembling two micro metal gear motors. Don't forget to solder the wires to each motor terminal and add **0.1 µF ceramic capacitors** between them for noise reduction (https://www.pololu.com/docs/0J15/9). After building the chassis, you will get the Zumo robot as shown in *Figure 5.2*:

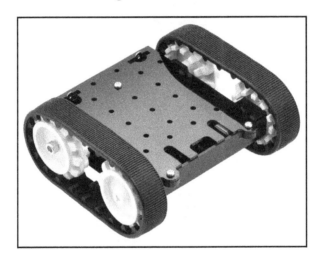

Figure 5.2: An assembled Zumo chassis kit. Image courtesy of Pololu (https://www.pololu.com/)

There are two solder tabs that protrude through the holes on the top of the chassis. Solder the wires to the tabs to get the power from the battery compartment. It can provide approximately 4.8V with fresh rechargeable NiMH AAA batteries.

Attaching the Feather Doubler

The FeatherWing Doubler (https://www.adafruit.com/product/2890) allows you to connect together the Feather HUZZAH ESP8266 board and the DC Motor + Stepper FeatherWing Add-on board. *Figure 5.3* shows a FeatherWing Doubler with soldered Feather stacking headers:

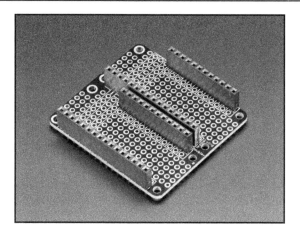

Figure 5.3: A FeatherWing Doubler. Image courtesy of Adafruit Industries (https://www.adafruit.com/)

1. Stick four bumper feet near the corners of the FeatherWing Doubler, but don't cover up any of the mounting holes (*Figure 5.4*):

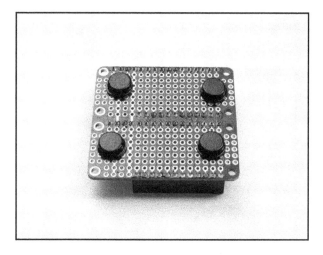

Figure 5.4: Sticking the bumper feets https://learn.adafruit.com/assets/29484. Image courtesy of Adafruit Industries (https://www.adafruit.com/)

2. Now, attach the FeatherWing Doubler to the black acrylic mounting plate of the chassis using two 4-40 screws and nuts.
3. Place the Feather HUZZAH ESP8266 on the FeatherWing Doubler.

4. Take a stick of extra-long male header and break off five pieces, carefully pulling or cutting out the middle one (*Figure 5.5*):

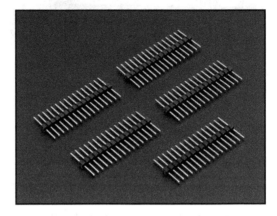

Figure 5.5: Extra-long break-away 0.1" 16-pin strip male header (five pieces). Image courtesy of Adafruit Industries (https://www.adafruit.com/)

5. Insert the prepared extra-long male header into M3 and M4. Then tighten the screws in terminal blocks. This will make it easier to connect and disconnect the DC motors from the DC Motor + Stepper FeatherWing Add-on board (*Figure 5.6*):

Figure 5.6: Inserting the male headers into M3 and M4. Image courtesy of Adafruit Industries (https://www.adafruit.com/)

6. Connect leads from the left motor to the M3 connector block of the DC Motor + Stepper FeatherWing Add-on board.

7. Connect leads from the right motor to M4 of the DC Motor + Stepper FeatherWing Add-on board.

8. Place the DC Motor + Stepper FeatherWing Add-on board next to the Feather HUZZAH ESP8266 on the FeatherWing Doubler.

9. Connect the power leads from the 4 x AA battery pack to the power connector of the motor driver. Make sure that you turn off the switch before connecting to the motor driver. The battery pack provides about 4.8V from fresh NiMH batteries. Each battery provides 1.2V and the capacity is 2200 mAh.

10. Using a piece of double-sided sticky tape, attach the Li-Poly battery to the Zumo chassis close to the Feather HUZZAH ESP8266. Then, connect the leads from the battery to the battery connector of ESP8266.

11. Connect the SparkFun Line Follower Array to the front of the Zumo chassis as shown in *Figure 5.7*:

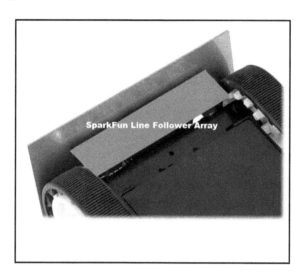

Figure 5.7: Connecting the SparkFun Line Follower Array. Image (Edited) courtesy of Plololu (https://www.pololu.com/)

Wire the SparkFun Line Follower Array with the Feather HUZZAH ESP8266 as presented in *Table 5-1*:

SparkFun Line Follower Array	Feather HUZZAH ESP8266
Power - 5V DC	3V
Ground	GND
I2C Data	SDA/4
I2C Clock	SCL/5

Table 5-1: Wiring connections between the Line Follower Array and the Feather HUZZAH ESP8266

Figure 5.8 shows the wiring between the SparkFun Line Follower Array and the Feather HUZZAH ESP8266:

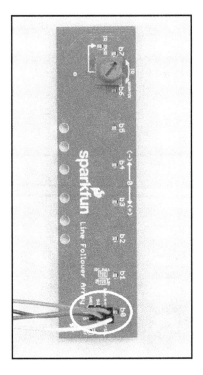

Figure 5.8: Connecting the SparkFun Line Follower Array with the Feather HUZZAH ESP8266. Image (Edited) courtesy of SparkFun Electronics (https://www.sparkfun.com/)

Now you have completely assembled the Zumo chassis kit with all the required components to build a line-following robot.

A line-following robot requires a line-following course (track) to drive on. It could be a dark track on a light background or a light track on a dark background. In the next section, you will learn how to build a line-following course on a light background.

Building a line-following course

The following steps will show you how to build a line-following course.

You will need the following things to build the course:

- Boxboard
- Pencil
- Ruler
- Black permanent marker pen

Draw a sketch of your course using a pencil and ruler. The width of the line should be approximately 1" and the course should be connected (not open) to drive the robot continuously. Then, color the track using a permanent marker pen. *Figure 5.9* shows a sample line-following course that you can use to build one:

Figure 5.9: A Line-following course. Image courtesy of Pololu (https://www.pololu.com/)

Writing Arduino sketch

You will need the following Arduino libraries to write the Arduino sketch to control two motors for detecting the track:

- SparkFun line follower Array Arduino library

- Adafruit Motor Shield V2 library: Download it from
  `https://github.com/ladyada/Adafruit_Motor_Shield_V2_Library/archive/master.zip`

Listing 5-1 – Arduino sketch for line-following

Listing 5-1 shows a snippet from the complete Arduino sketch that you can use to upload to the Feather HUZZAH ESP8266 board. The complete code can be found in the example code file of this chapter:

```
#include "ESP8266WiFi.h"
#include <Wire.h>
#include <Adafruit_MotorShield.h>
#include "sensorbar.h" //SparkFun line follower array Arduino library

// Create the motor shield object with the default I2C address
Adafruit_MotorShield AFMS = Adafruit_MotorShield();

// And connect 2 DC motors to port M3 & M4 !
Adafruit_DCMotor *L_MOTOR = AFMS.getMotor(4);
Adafruit_DCMotor *R_MOTOR = AFMS.getMotor(3);

// Uncomment one of the four lines to match your SX1509's address
//  pin selects. SX1509 breakout defaults to [0:0] (0x3E).
const uint8_t SX1509_ADDRESS = 0x3E;  // SX1509 I2C address (00)
//const byte SX1509_ADDRESS = 0x3F;  // SX1509 I2C address (01)
//const byte SX1509_ADDRESS = 0x70;  // SX1509 I2C address (10)
//const byte SX1509_ADDRESS = 0x71;  // SX1509 I2C address (11)

SensorBar mySensorBar(SX1509_ADDRESS);

//Define motor polarity for left and right side -- USE TO FLIP motor
directions if wired backwards
#define RIGHT_WHEEL_POL 1
#define LEFT_WHEEL_POL 1

//Define the states that the decision-making machines uses:
#define IDLE_STATE 0
#define READ_LINE 1
#define GO_FORWARD 2
#define GO_LEFT 3
#define GO_RIGHT 4

uint8_t state;
...
```

The following steps will discuss some of the most important configurations and functions of the Arduino sketch:

1. First, include all the required libraries:

```
#include "ESP8266WiFi.h"
#include <Wire.h>
#include <Adafruit_MotorShield.h>
#include "sensorbar.h" //SparkFun line follower array Arduino
library
```

2. After this, you create an instance of `Adafruit_MotorShield` and also create instances for the two motors:

```
// Create the motor shield object with the default I2C address
Adafruit_MotorShield AFMS = Adafruit_MotorShield();

// And connect 2 DC motors to port M3 & M4 !
Adafruit_DCMotor *L_MOTOR = AFMS.getMotor(4);
Adafruit_DCMotor *R_MOTOR = AFMS.getMotor(3);
```

3. Then, you create an instance of `SensorBar` and connect it to the I2C bus with the address `0x3E`. If you need to change the address of the array, move the solder jumper to set `A0` and `A1`. The silkscreen table gives a reference. The default address is `0x3E` (*Figure 5.10*). For example, if you want to use address `0x3F`, move `A0` to the `1` position and leave `A1` at `0`:

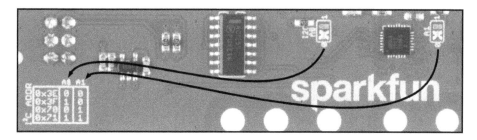

Figure 5.10: Selecting the I2C address. Image courtesy of SparkFun Electronics (https://www.sparkfun.com)

```
// Uncomment one of the four lines to match your SX1509's
address
//  pin selects. SX1509 breakout defaults to [0:0] (0x3E).
const uint8_t SX1509_ADDRESS = 0x3E;  // SX1509 I2C address
(00)
//const byte SX1509_ADDRESS = 0x3F;  // SX1509 I2C address (01)
//const byte SX1509_ADDRESS = 0x70;  // SX1509 I2C address (10)
//const byte SX1509_ADDRESS = 0x71;  // SX1509 I2C address (11)

SensorBar mySensorBar(SX1509_ADDRESS);
```

4. Now, add the following constants, variables, and function prototypes:

```
//Define motor polarity for left and right side -- USE TO FLIP
motor directions if wired backwards
#define RIGHT_WHEEL_POL 1
#define LEFT_WHEEL_POL 1

//Define the states that the decision-making machines uses:
#define IDLE_STATE 0
#define READ_LINE 1
#define GO_FORWARD 2
#define GO_LEFT 3
#define GO_RIGHT 4

uint8_t state;
void driveBot( int16_t driveInput );
void turnBot( float turnInput );
void stop();
```

5. Inside the `setup()` function of the sketch, initialise the `Adafruit_MotorShield` library:

```
AFMS.begin();
```

6. Turn on the sensor bar's IR strobing:

```
mySensorBar.setBarStrobe();
```

7. Add the following line to detect the dark line on the light background:

```
// dark line on light background
mySensorBar.clearInvertBits();
```

8. Now, start the sensor bar's I2C expander with the following command:

```
mySensorBar.begin();
```

9. Inside the `loop()` function, read the sensor position on the dark line (*Figure 5.11*) with the following command:

```
int position = mySensorBar.getPosition();
```

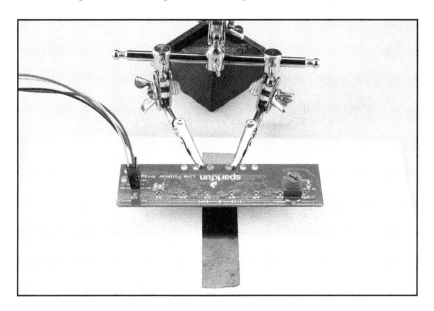

Figure 5.11: A sensor placed on a dark line. Only b3 and b4 IR sensors can detect the line. Image courtesy of SparkFun Electronics (https://www.sparkfun.com)

10. Define a variable to store the next state:

```
uint8_t nextState = state;
```

11. The following switch statement allows you to control the motors according to the value held in the state variable. To drive the robot parallel to the dark line, the sensor should output values between -50 to +50.
If the sensor value is less than -50, it means that the robot is turning RIGHT. So, the program should send commands to the motors to turn the robot to the LEFT to correct the movement.

If the sensor value is greater than 50, it means that the robot is turning LEFT. So, the program should send commands to the motors to turn the robot to the RIGHT to correct the movement.

The preceding two mechanisms will help drive the robot more parallel to the dark line:

```
switch (state) {
case IDLE_STATE:
   motors.stop();          // Stops both motors
   nextState = READ_LINE;
   break;
  case READ_LINE:
   if( mySensorBar.getDensity() < 7 )
   {
     nextState = GO_FORWARD;
     if( mySensorBar.getPosition() < -50 )
     {
       nextState = GO_LEFT;
     }
     if( mySensorBar.getPosition() > 50 )
     {
       nextState = GO_RIGHT;
     }
   }
   else
   {
     nextState = IDLE_STATE;
   }
   break;
  case GO_FORWARD:
   driveBot(200);
   nextState = READ_LINE;
   break;
  case GO_LEFT:
   turnBot(-.75);
   nextState = READ_LINE;
   break;
  case GO_RIGHT:
   turnBot(.75);
   nextState = READ_LINE;
   break;
  default:
stop();          // Stops both motors
   break;
  }
  state = nextState;
  //delay(100);
}
```

12. The following function will drive the robot straight on the dark line with the given speed between -255 to 255. Use a positive number to drive forward and a negative number to drive backward:

```
void driveBot( int16_t driveInput ) {
    int16_t rightVar;
    int16_t leftVar;
    rightVar = driveInput * RIGHT_WHEEL_POL;
    leftVar = -1 * driveInput * LEFT_WHEEL_POL;
    R_MOTOR->setSpeed(rightVar);
L_MOTOR->setSpeed(leftVar);
}
```

13. The following function will turn the robot to the left or right. turnInput can be between -1 to +1. Use negative values for turning left and positive values for turning right. Use the positive value 1 to spin right at the maximum speed:

```
void turnBot( float turnInput )
{
    int16_t rightVar;
    int16_t leftVar;
    //If turn is positive
    if( turnInput > 0 )
    {
        rightVar = -1 * 255 * RIGHT_WHEEL_POL * turnInput;
        leftVar = -1 * 255 * LEFT_WHEEL_POL * turnInput;
    }
    else
    {
        rightVar = 255 * RIGHT_WHEEL_POL * turnInput * -1;
        leftVar = 255 * LEFT_WHEEL_POL * turnInput * -1;
    }

    R_MOTOR->setSpeed(rightVar);
L_MOTOR->setSpeed(leftVar);
    delay(5);
}
```

14. The following function will stop both motors:

```
void stop() {

  L_MOTOR->setSpeed(0);
  L_MOTOR->run( RELEASE );

  R_MOTOR->setSpeed(0);
  R_MOTOR->run( RELEASE );
```

```
    }
```

Uploading the sketch

Now, it's time to upload the sketch to the Feather HUZZAH ESP8266 using the Arduino IDE:

- If you're using the Windows operating system, first install the SiLabs CP2104 driver. The driver can be downloaded from `http://www.silabs.com/products/development-tools/software/usb-to-uart-bridge-vcp-drivers`.
- If you are using Mac OS 10.12.6 (Sierra) and you cannot upload with the latest Mac OS VCP driver, please try the legacy v4 driver. Note that you will need to uninstall the v5 driver using uninstall.sh (in the driver package). The driver can be downloaded from `http://community.silabs.com/t5/Interface-Knowledge-Base/Legacy-OS-Software-and-Driver-Packages/ta-p/182585`.

The following steps will guide you how to configure and upload the sketch to the Feather HUZZAH ESP8266:

1. Install the ESP8266 board package on Arduino IDE if you haven't installed it in `Chapter 1`, *Getting Ready*.
2. Restart the Arduino IDE to use new the libraries and drivers.
3. Select the Adafruit HUZZAH ESP8266 from the **Tools | Board** drop-down.
4. Select **80 MHz** as the CPU frequency.
5. Select **115200 baud** as the upload speed.
6. Select the matching COM port associated with HUZZAH ESP8266.
7. Verify the Arduino sketch by clicking on the **Verify** button in the toolbar.
8. Click on the **Upload** button to upload the sketch.

Playing with your robot

After uploading the Arduino sketch, you are ready to play with the Zumo robot:

1. First, insert the batteries in the battery compartment of the Zumo robot.

2. Then, place the robot on the dark line of the race track and turn on the power button. The robot will start to follow the dark line (*Figure 5.12*). Line-following robots are very susceptible to light and ground texture changes and may show unexpected line-following behaviors. If so, adjust the contrast and lighting on the environment:

Figure 5.12: A line-following Zumo robot. Image courtesy of Pololu (https://www.pololu.com)

Summary

In this chapter, you built a line-following Zumo robot with the Feather HUZZAH ESP8266 to drive on a dark line on a light background.

In Chapter 6, *Building an ESP8266 Robot Controller*, you will build an ESP8266-based robot that can be controlled using a Wi-Fi remote.

6

Building an ESP8266 Robot Controller

This chapter will guide you to build an ESP8266-based simple Robot Controller that can be used to control the movement of the Romi Robot Chassis introduced under *building the Romi Robot* section. The Robot Controller consists of four movement-related push buttons that can be used to control the movement of the Romi Robot in four directions: forward, backward, left, and right. A Wi-Fi network is used to connect the Robot Controller with the Romi Robot, and a Blynk bridge is used to connect both of them for handling bi-directional data communication. This Robot Controller is very similar to a RC Remote Controller (*Figure 6.1*) but uses a Wi-Fi network to communicate with the robot instead of radio frequency:

Figure 6.1: A commercial RC Car Remote Controller that can be used to control a hobby car in four directions

Let's get started with building the Robot Controller and the Romi Robot Chassis.

In this chapter, you will learn the following topics:

- Assembling the Romi Robot Chassis
- Connecting electronics to the Romi Robot Chassis
- Building a Robot Controller
- Writing Arduino sketches for the Romi Robot and the Robot Controller with Blynk Bridge

Things you will need

First, let's prepare with all the things that you will need to build the Romi Robot and the Robot Controller. Here is the list of things:

- 1 x Romi Chassis Kit (`https://www.pololu.com/category/203/%20romi-chassis-kits`)
- 2 x Adafruit Feather HUZZAH ESP8266 (`https://www.adafruit.com/product/3046`)
- 1 x DC Motor + Stepper FeatherWing Add-on (`https://www.adafruit.com/product/2927`)
- 1 x FeatherWing Doubler (`https://www.adafruit.com/product/2890`)
- 1 x Extra-long break-away 0.1" 16-pin strip male header (`https://www.adafruit.com/product/400`)
- 4 x Push buttons (`https://www.sparkfun.com/`)
- 2 x 2000mAh 3.7 Li-Poly battery (`https://www.adafruit.com/product/2011`)
- 1 x Breadboard
- 4 x Rechargeable NiMH AA batteries—1.2V, 2200 mAh (`https://www.pololu.com/product/1003`)

Building the Romi Robot

The Romi chassis (*Figure 6.2*) is a differential-drive mobile robot platform with a diameter of 6.5" (165 mm) and an integrated battery holder for six AA batteries. The drive wheels are located on a diameter of the circular base plate, allowing for turns or spinning in place. A large, fixed ball caster in the rear provides a smooth third point of contact and a second ball caster can be optionally added to the front.

The Romi chassis has plenty of general-purpose mounting holes and slots intended for various screw sizes, such as #2-56, #4-40, M2, and M3. These holes can be used for mounting your electronics, such as your microcontroller, motor drivers, and sensors:

Figure 6.2: The Romi Chassis Kit -Black. Image courtesy of Pololu—https://www.pololu.com

The Romi chassis kit (*Figure 6.2*) comes with the following things and requires assembly:

- One black Romi chassis base plate with battery cover
- Two mini plastic gearmotors (120:1 HP with offset output and extended motor shaft)
- A pair of black Romi chassis motor clips
- A pair of white 70 × 8 mm Pololu wheels
- One black Romi chassis ball caster kit:
    - One ball caster retention clip
    - One 1" plastic ball
    - Three rollers for ball caster
- One Romi chassis battery contact set:
    - Two battery contacts, double
    - Two battery contact terminals, spring
    - Two battery contact terminals, flat

Figure 6.3: Contents of the Romi chassis kit. Image courtesy of Pololu—https://www.pololu.com

Additionally, you will need six rechargeable NiMH AA batteries (*Figure 6.4*) with the capacity of 2200 mAh each. Remember that you will need to charge these batteries before you use them:

Figure 6.4: Rechargeable NiMH AA Battery—1.2V, 2200 mAh. Image courtesy of Pololu—https://www.pololu.com

Ball casters

The ball caster provides a smooth rolling action to the Romi Robot Chassis. First, install the ball caster onto the Romi Robot Chassis. The following steps will describe how to install it with the provided supporting components. With the chassis upside down, perform the following steps:

1. Place the three rollers for the rear ball caster into the cut-outs in the chassis (*Figure 6.5*):

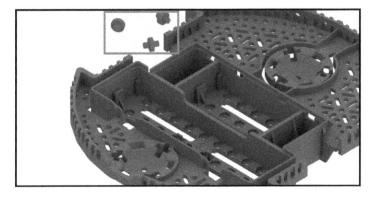

Figure 6.5: Placing the three rollers. Image courtesy of Pololu—https://www.pololu.com

2. Place the 1" plastic ball on top of the three rollers (*Figure 6.6*):

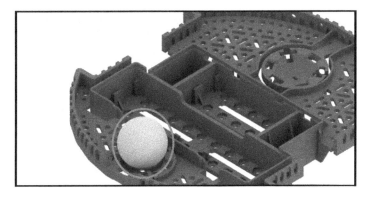

Figure 6.6: Placing the plastic ball. Image courtesy of Pololu—https://www.pololu.com

3. Push the ball caster retention clip over the ball and into the chassis so that the three legs snap into their respective holes (*Figure 6.7*):

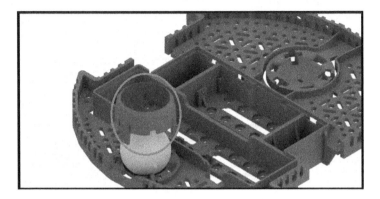

Figure 6.7: Connecting the ball caster retention clip. Image courtesy of Pololu—https://www.pololu.com

Battery contacts

Before inserting batteries in the battery box of the Romi Robot Chassis, you should install the battery contacts to the slots. The following steps will describe how to install them:

1. Push the two double-sided battery contacts to the slots (*Figure 6.8*):

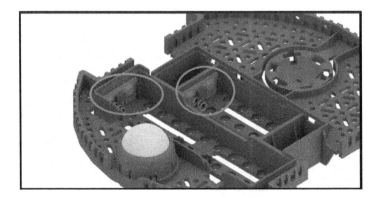

Figure 6.8: Connecting the double-sided battery contacts. Image courtesy of Pololu—https://www.pololu.com

2. Turn the chassis over, and place the four individual battery contact terminals into the chassis from the top of the battery box (*Figure 6.9*):

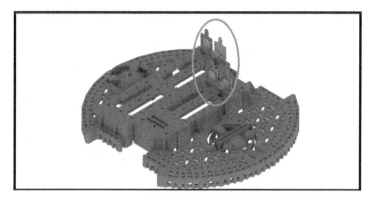

Figure 6.9: Connecting the individual battery contact terminals. Image courtesy of Pololu—https://www.pololu.com

3. After placing the battery contact terminals, you can use the four terminal tabs to get power from the batteries (*Figure 6.10*):

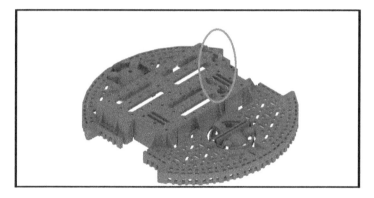

Figure 6.10: Tabs from the battery contact terminals. Image courtesy of Pololu—https://www.pololu.com

Motors

The following steps will describe how to install motor clips and mini plastic gear motors to the Romi Robot Chassis:

1. Align the motor clip with the chassis as shown in *Figure 6.11*:

Figure 6.11: Aligning the motor clip. Image courtesy of Pololu—https://www.pololu.com

2. Press the clip firmly into the chassis until the bottom of the clip is even with the bottom of the chassis. You may hear several clicks when you are pressing the motor clip (*Figure 6.12*):

Figure 6.12: The motor clip connected to the base plate. Image courtesy of Pololu—https://www.pololu.com

3. Push the mini plastic gearmotor into the motor clip until it snaps into place (*Figure 6.13*). You may hear a click!

Figure 6.13: Pushing the mini plastic gearmotor into the motor clip. Image courtesy of Pololu—https://www.pololu.com

Wheels

After installing the mini plastic gear motors, it's time to attach wheels onto them. The following steps will describe how to attach the wheels onto the output shaft of the mini plastic gear motors:

1. Align the wheel with the output shaft of the motor (*Figure 6.14*):

Figure 6.14: Aligning the wheel with the output shaft of the motor. Image courtesy of Pololu—https://www.pololu.com

2. Press the wheel onto the output shaft of the motor until the motor shaft is flush with the outer face of the wheel (*Figure 6.15*):

Figure 6.15: Pressing the wheel onto the output shaft of the motor. Image courtesy of Polol—https://www.pololu.com

3. Repeat the preceding steps to assemble the other wheel on the base plate with the motor clip and the gear motor (*Figure 6.16*):

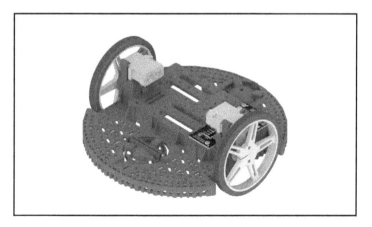

Figure 6.16: The completed wheel assembly. Image courtesy of Pololu—https://www.pololu.com

4. Solder the wires to the mini plastic gear motors.

The battery compartment and power distribution

The battery compartment is separated into two sections: a four-battery section and a two-battery section. Either section of the battery compartment can be used independently. The Adafruit Stepper + DC Motor FeatherWing requires about 4.5V to drive two mini plastic gearmotors. However, they should run comfortably in the 3V to 6V range.

The four-battery section can produce 4.8V with fully recharged 1.2V NiMH AA batteries. This voltage is more than enough to power the motors through the Stepper + DC Motor FeatherWing.

Solder two wires (red and black) to the positive and negative tabs of the four-battery section of the battery compartment (*Figure 6.17*):

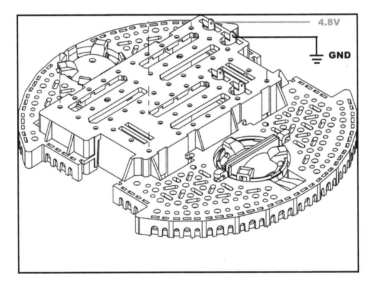

Figure 6.17: Soldering wires to the four-battery section. Image (edited) courtesy of Pololu—https://www.pololu.com

Attaching the FeatherWing Doubler

Now, you're going to attach the FeatherWing Doubler to the Romi Robot Chassis. The FeatherWing Doubler provides mechanism to connect the Feather HUZZAH ESP8266 and the Stepper + DC Motor FeatherWing together:

1. Using four aluminum standoffs, attach the FeatherWing Doubler - Prototyping Add-on to the base plate (*Figure 6.18*):

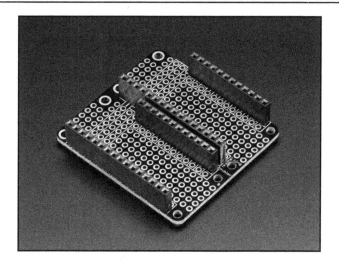

Figure 6.18: Adafruit FeatherWing Doubler - Prototyping Add-on. Image courtesy of Adafruit Industries—https://www.adafruit.com

2. Place the Adafruit Feather HUZZAH ESP8266 on the FeatherWing Doubler as shown in *Figure 6.19*:

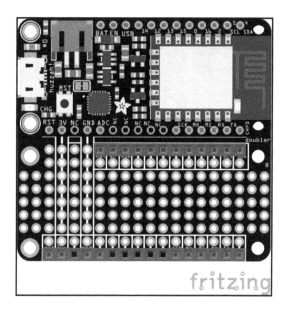

Figure 6.19: Placing the Adafruit Feather HUZZAH ESP8266 on the FeatherWing Doubler

3. Take the Adafruit Stepper + DC Motor FeatherWing and a stick of extra-long male header and break off five pieces, carefully pulling (or cutting) out the middle one.

4. Insert it into M3 and M4. Secure it by tighten the screws on terminal blocks. This will make it easy to connect and disconnect the DC motors (*Figure 6.20*):

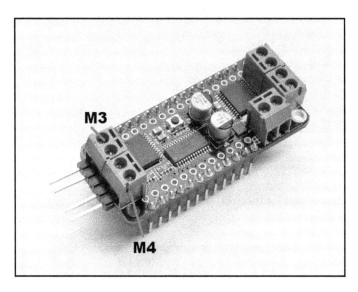

Figure 6.20: The Adafruit Stepper + DC Motor FeatherWing prepared with extra-long male headers. Image courtesy of Adafruit Industries—https://www.adafruit.com

5. Place the Adafruit Stepper + DC Motor FeatherWing on the FearherWing Doubler next to the Feather HUZZAH ESP8266 (*Figure 6.21*):

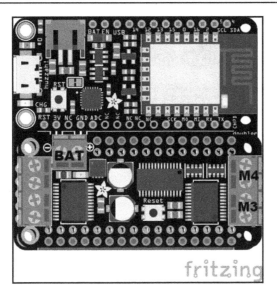

Figure 6.21: Placing the DC Motor FeatherWing on the FeatherWing Doubler

6. Connect the wires from the gear motors with the DC Motor FeartherWing. Connect the left motor with the M4 terminal block and the right motor with the M3 terminal block (*Figure 6.22*):

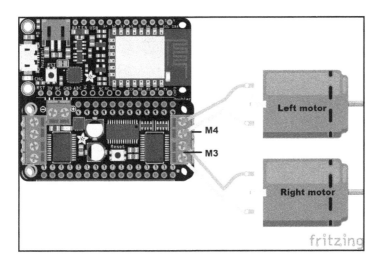

Figure 6.22: Wiring the gear motors with the DC Motor FeatherWing

7. Connect the + and - lines from the battery compartment to the DC Motor FeatherWing power terminals (*Figure 6.23*). Don't insert the batteries in the battery compartment yet:

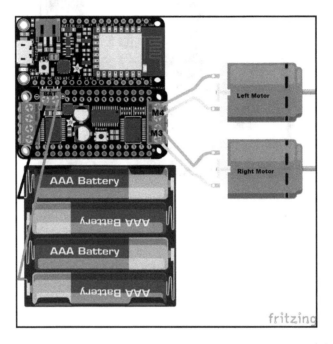

Figure 6.23: Powering the DC Motor FeatherWing with the four battery section of the battery compartment

Building the Robot Controller

Let's get started with building the Robot Controller that can be used to control the Romi Robot over the Wi-Fi network. The controller has four movement related push buttons that can be used to control the movement of the robot (*Figure 6.24*). The following is a list of operations associated with each movement-related push button:

- **A**: FORWARD
- **B**: BACKWARD
- **C**: LEFT
- **D**: RIGHT

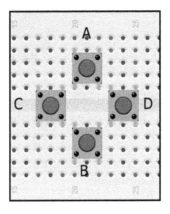

Figure 6.24: The Robot Controller movement-related push buttons

Figure 6.25 shows the wiring diagram that can be used to build the Robot Controller with the Adafruit Feather HUZZAH ESP8266 and the four movement-related push buttons. You can use any GPIO pin (#0, #2, #4, #5, #12, #13, #14, #15, or #16) at the top edge of the Feather PCB to connect to the push buttons. The wiring diagram uses GPIO #12, #13, #14, and #15 as the input. Connect the push buttons with GPIO pins, as follows:

- **Push button A (FORWARD)**: GPIO #12
- **Push button B (BACKWARD)**: GPIO #13
- **Push button C (LEFT)**: GPIO #14
- **Push button D (RIGHT)**: GPIO #15

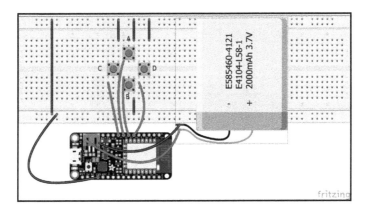

Figure 6.25: Wiring diagram for the Robot Controller

Software

In this project, you will use a Blynk Bridge to connect the Robot Controller and the Romi Robot through a Wi-Fi network to handle all the data communication between them.

The Bridge can be used for Device-to-device communication (no app. involved). With a Blynk bridge, you can send digital, analog, or virtual write commands from one device to another by knowing its Auth Token:

- First, create two Blynk projects for the button controller and the robot:
  - **Project 1**: Controller
  - **Project 2**: Robot

- Get the API key for both projects. You need Auth keys to connect both devices through the Blynk Bridge. However, you can write sketches without adding the Blynk Bridge widget to your app.

Arduino sketch for the Robot Controller

Listing 6-1 shows the Arduino sketch that you can use to upload to the Feather HUZZAH ESP8266 residing in the Robot Controller.

Listing 6-1—Arduino sketch for Controller is as follows:

```
#include <ESP8266WiFi.h>
#include <BlynkSimpleEsp8266.h>

//Buttons
const int forward = 12;
const int backward = 13;
const int left = 14;
const int right = 15;

// You should get Auth Token in the Blynk App.
// Go to the Project Settings (nut icon).
char auth[] = "YourAuthToken";

// Your WiFi credentials.
// Set password to "" for open networks.
char ssid[] = "XXXXXXXX";
char pass[] = "********";

WidgetBridge bridge1(V1); //Initiating Bridge Widget on V1 of CONTROLLER
```

```
void setup() {

  // initialize the pushbutton pin as an input:
  pinMode(forward, INPUT);
  pinMode(backward, INPUT);
  pinMode(left, INPUT);
  pinMode(right, INPUT);

  Blynk.begin(auth, ssid, pass);
  while (Blynk.connect() == false) {
    // Wait until Blynk is connected
  }

}

BLYNK_CONNECTED() {
  bridge1.setAuthToken("OtherAuthToken"); // Token for the ROBOT
}

void loop() {
  Blynk.run();

  // read the state of the pushbutton FORWARD:
  int forwardButtonState = digitalRead(forward);

  // check if the pushbutton is pressed. If it is, the forwardButtonState
is HIGH:
  if (forwardButtonState == HIGH) {
    bridge1.virtualWrite(V1, 1);
  } else {
    bridge1.virtualWrite (V1, 0);
  }

  // read the state of the pushbutton BACKWARD:
  int backwardButtonState = digitalRead(backward);

  // check if the pushbutton is pressed. If it is, the backwardButtonState
is HIGH:
  if (backwardButtonState == HIGH) {
    bridge1.virtualWrite(V1, 2);
  } else {
    bridge1.virtualWrite (V1, 0);
  }

  // read the state of the pushbutton LEFT:
  int leftButtonState = digitalRead(left);

  // check if the pushbutton is pressed. If it is, the leftButtonState is
```

```
HIGH:
  if (leftButtonState == HIGH) {
    bridge1.virtualWrite(V1, 3);
  } else {
    bridge1.virtualWrite (V1, 0);
  }

  // read the state of the pushbutton RIGHT:
  int rightButtonState = digitalRead(right);

  // check if the pushbutton is pressed. If it is, the rightButtonState is
HIGH:
  if (rightButtonState == HIGH) {
    bridge1.virtualWrite(V1, 4);
  } else {
    bridge1.virtualWrite (V1, 0);
  }

}
```

Type the preceding sketch using your Arduino IDE. Then, select **Tools** I **Board** and then Adafruit HUZZAH ESP8266 from the Arduino IDE's toolbar. After this, connect the Feather HUZZAH ESP8266 to the computer using a micro-USB cable. After flashing the sketch, remove the micro-USB cable from the Feather HUZZAH ESP8266. Now, you can use the Robot Controller as a hand-held device with a 3.7V Li-Poly battery.

Coding the Romi Robot

You can connect the Romi Robot with the Robot Controller via the same Blynk Bridge.

Listing 6-2 – Arduino sketch for the Robot

Listing 6-2 shows the Arduino sketch that can be flashed to the Romi Robot to read the incoming data from the Robot Controller and control the two gear motors accordingly:

```
#include <ESP8266WiFi.h>
#include <BlynkSimpleEsp8266.h>
#include <Wire.h>
#include <Adafruit_MotorShield.h>

// You should get Auth Token in the Blynk App.
// Go to the Project Settings (nut icon).
char auth[] = "xxxxxxxxxxxxxxxxxxxxxxxxxxxxxx";
```

```
// Your WiFi credentials.
// Set password to "" for open networks.
char ssid[] = "XXXXXXXX";
char pass[] = "********";

Adafruit_MotorShield AFMS = Adafruit_MotorShield();

Adafruit_DCMotor *L_MOTOR = AFMS.getMotor(4);
Adafruit_DCMotor *R_MOTOR = AFMS.getMotor(3);

// This code will update the virtual port 1
BLYNK_WRITE(V1) {
  int pinData = param.asInt();
  if (pinData == 1){
    forward();
  }
  else if (pinData == 2){
    backward();
  }
  else if (pinData == 3){
    left();
  }
  else if (pinData == 4){
    right();
  }
  else {
    stop();
  }
}

void setup(){
  //Serial.begin(9600);
  Blynk.begin(auth, ssid, pass);

  AFMS.begin();
}

void loop(){
  Blynk.run();
}

void forward() {

  L_MOTOR->setSpeed(200);
  L_MOTOR->run( FORWARD );

  R_MOTOR->setSpeed(200);
  R_MOTOR->run( FORWARD );
```

```
  }
  void backward() {

    L_MOTOR->setSpeed(150);
    L_MOTOR->run( BACKWARD );

    R_MOTOR->setSpeed(150);
    R_MOTOR->run( BACKWARD );

  }

  void stop() {

    L_MOTOR->setSpeed(0);
    L_MOTOR->run( RELEASE );

    R_MOTOR->setSpeed(0);
    R_MOTOR->run( RELEASE );

  }

  void left() {

    L_MOTOR->setSpeed(100);
    L_MOTOR->run( BACKWARD );

    R_MOTOR->setSpeed(100);
    R_MOTOR->run( FORWARD );

  }

  void right() {

    L_MOTOR->setSpeed(100);
    L_MOTOR->run( FORWARD );

    R_MOTOR->setSpeed(100);
    R_MOTOR->run( BACKWARD );

  }
```

After flashing the code to the Feather HUZZAH ESP8266 attached to the Romi Robot, remove the micro-USB cable.

Then, connect the 3.7V 2000 mAh Li-Poly battery to the battery connector of the Feather HUZZAH ESP8266 (*Figure 6.26*):

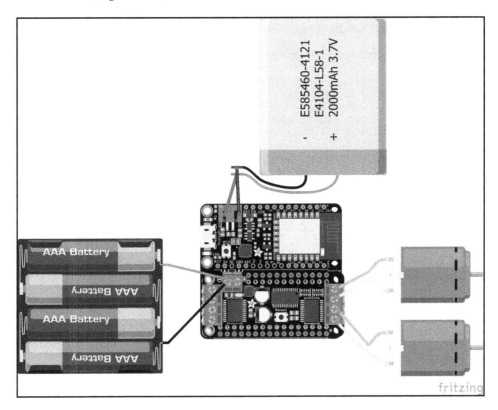

Figure 6.26: Connecting the Li-Poly batteries to the Feather HUZZAH ESP8266

Play it

First, insert four rechargeable AA NiMH batteries in the four battery section of the battery compartment. Then, insert the battery cover to complete the assembly (*Figure 6.27*):

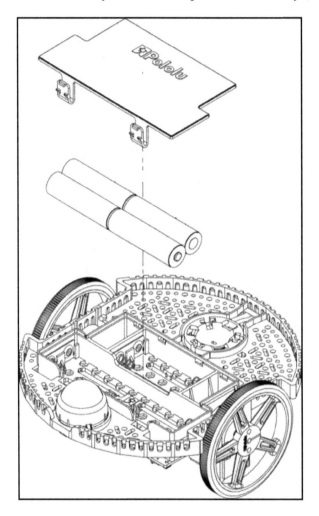

Figure 6.27: Inserting the batteries and the battery cover. Image (edited) courtesy of Pololu—https://www.pololu.com

Now you're ready to control the Romi Robot with the hand-held Robot Controller.

Press and hold the push button on the Robot Controller to move the robot in four directions:

- FORWARD
- BACKWARD
- LEFT
- RIGHT

Of course, the DC Motor driver doesn't actually know if the gear motor is FORWARD or BACKWARD, so if you want to change which way it thinks is forward, simply swap the two wires from the motor to the DC Motor driver.

Summary

In this chapter, you built a Romi Robot and a Robot Controller based on the Feather HUZZAH ESP8266. Then, you programmed both devices with Arduino sketches to control the Romi Robot, using the Robot Controller through a Wi-Fi network. You used a Blynk Bridge to connect both devices. Further, you learned to modify the Robot Controller by adding more buttons to control more functions on the robot.

In Chapter 7, *Building a Gripper Robot*, you will build a robot with a gripper that can be used to grab and release small objects.

7
Building a Gripper Robot

A gripper is just like a human hand. It enables the holding, tightening, handling, and releasing of an object. A gripper can be attached to the chassis of the robot and controlled with a microcontroller. Many styles and sizes of grippers exist so that the correct model can be selected for the application.

In this chapter, you will learn the following topics:

- Building a robot with a gripper using Mini Robot Rover Chassis kit
- Assembling a parallel gripper kit with a servo motor
- Building the circuit with ESP8266, Arduino, and the Motor/Stepper/Servo shield
- Writing an Arduino sketch to control the gripper through a Blynk app

Things you will need

The following is the list of things you will need to build the Gripper Robot:

- 1 x Mini Robot Rover Chassis Kit - 2WD with DC Motors (`https://www.adafruit.com/product/2939`)
- 1 x Parallel Gripper Kit A - Channel mount (`https://www.sparkfun.com/products/13178`)
- 1 x Servo - Hitec HS-422 (standard size) (`https://www.sparkfun.com/products/11884`)
- 1 x Arduino Uno R3 (Atmega328 - assembled)
- 1 x Adafruit Motor/Stepper/Servo Shield for Arduino v2 Kit - v2.3 (`https://www.adafruit.com/product/1438`)

- 1 x 9V battery holder with switch and 5.5mm/2.1mm plug (https://www.adafruit.com/product/67)
- 1 x alkaline 9V battery (https://www.adafruit.com/product/1321)
- 1 x 4 x AA battery holder with on/off switch (https://www.adafruit.com/product/830)
- 1 x alkaline AA battery (LR6) - 4 pack (https://www.adafruit.com/product/3349)

Mini Robot Rover chassis kit

The Mini Robot Rover chassis kit (*Figure 7.1*) gives you everything you need to build the chassis of a 2-wheel-drive Robot Rover. The kit comes with the following components and requires assembly to build the complete chassis:

- 2 x wheels
- 2 x DC motors in MicroServo shape
- 1 x support wheel
- 1 x metal chassis
- 1 x top metal plate with mounting hardware

Figure 7.1: Mini Robot Rover Chassis Kit - 2WD with DC Motors. Image courtesy Adafruit Industries (https://www.adafruit.com)

Assembling the chassis

The following instructions will guide you on how to assemble the Mini Robot Rover chassis with the supplied components:

- Flip over the metal chassis and attach the two DC motors with the screws. Make sure that the wires of the DC motors are facing toward the front of the chassis.
- Attach the support wheel (caster wheel) to the front of the chassis with screws. The front side of the chassis has four holes that you can use to attach the caster wheel.
- Flip over the chassis again and attach two wheels to the DC motors with screws.
- You can attach the top plate after assembling the electronics.

Assembling the Gripper Kit

The Parallel Gripper Kit (*Figure 7.2*) comes with all the hardware you need to assemble the gripper:

Figure 7.2: Parallel Gripper Kit A - Channel Mount. Image courtesy of SparkFun Electronics (https://www.sparkfun.com)

The following instructions will guide you on how to assemble the Parallel Gripper Kit with the standard Hitec HS-422 servo:

1. You will need the following tools to assemble the Parallel Gripper Kit with servo:
   - Phillips-head screwdriver
   - Pliers

2. Take the following things out from the Gripper Kit:
   - ABS plate
   - Two acetal spacers
   - Four 7/16 x 6-32" pan head screws

3. Take the Hitec HS-422 servo (*Figure 7.3*) and remove the circle horn by removing the screw using the Phillips-head screwdriver:

Figure 7.3: Servo - Hitec HS-422 (standard size). Image courtesy of SparkFun Electronics (https://www.sparkfun.com)

4. Place the acetal spacers between the ABS plate and servo. Then, attach the servo to the ABS plate using the 7/16 x 6-32" pan head screws (*Figure 7.4*):

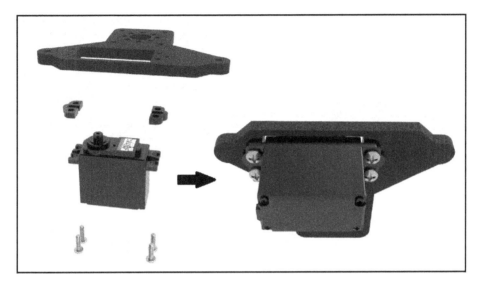

Figure 7.4: Attaching servo to the ABS base pulley. Image courtesy of ServoCity (https://www.servocity.com/parallel-gripper-kit-a)

5. Then, take the following things supplied with the kit to assemble a gripper to the right side of the ABS plate:

- Two 3/8 x 6-32" pan head screws
- Two acetal arms
- One ABS gripper
- Two 6/32 nylon nuts

6. Using the pan head screws and nylon nuts, connect the short arm of the acetal arm to the ABS plate and the long arm to the ABS gripper (*Figure 7.5*):

Figure 7.5: Assembling the right-side gripper. Image courtesy of ServoCity (https://www.servocity.com/parallel-gripper-kit-a)

7. Then, assemble the left-side gripper as shown in *Figure 7.6*:

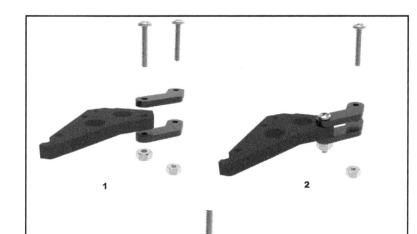

Figure 7.6: Assembling the left-side gripper. Image courtesy of ServoCity (https://www.servocity.com/parallel-gripper-kit-a)

8. Now, you need to take the following things:
   - Three 3/8 x 6-32" pan head screws
   - The servo screw and washer included with the servo
   - Two acetal gear arms

9. Connect the right-side acetal gear arm as shown in *Figure 7.7*. The gear end of the acetal gear arm should be connected to the shaft of the servo with the washer and screw that you removed earlier:

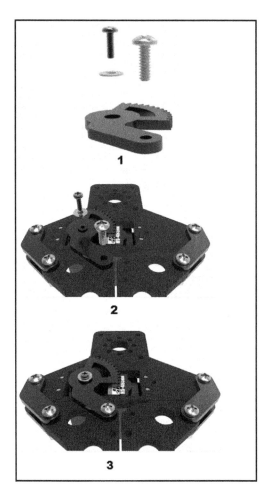

Figure 7.7: Assembling the right-side acetal gear arm. Image courtesy of ServoCity (https://www.servocity.com/parallel-gripper-kit-a)

10. Connect the left-side acetal gear arm, as shown in the *Figure 7.8*, using the screws that are provided:

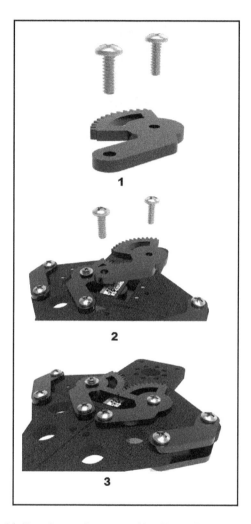

Figure 7.8: Assembling the left-side acetal gear arm. Image courtesy of ServoCity (https://www.servocity.com/parallel-gripper-kit-a)

11. After completing the assembly, your gripper arm should look similar to *Figure 7.9*:

Figure 7.9: The assembled gripper arm with servo motor. Image courtesy of ServoCity (https://www.servocity.com/parallel-gripper-kit-a)

Connecting the gripper to the chassis

The gripper can be easily mounted to the metal chassis of the Mini Robot Rover using two screws and nuts (*Figure 7.10*):

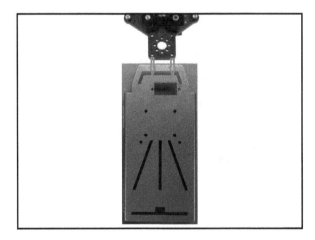

Figure 7.10: Connecting the gripper to the chassis

Assembling electronics

Follow the instructions given here to assemble the electronics on the Mini Robot Rover chassis. Use *Figure 7.11* to make connections between all the electronic components:

1. Take the Arduino UNO and mount it on the metal plate using the #4-40 machine screws and nuts. Use a piece of cardboard between the metal chassis and the Arduino to prevent the bottom of the Arduino from touching the metal robot chassis and shorting out.
2. Take the Adafruit Motor/Stepper/Servo Shield for Arduino v2 Kit - v2.3 and stack it on the Arduino UNO.
3. Connect the DC motors to the motor shield. The left DC motor should be connected to port M1 of the Motor Shield. The right DC motor should be connected to port M2 of the Motor Shield.
4. Screw the top plate standoffs into the mounting holes near the center of the chassis.
5. Screw the top plate firmly to the standoffs in the center.
6. Use double-sided foam tape to attach the 9V battery holder and the 4AA battery box to the top plate. Make sure to slide the switch to the **OFF** position for both 9V and 4xAA battery boxes.]
7. Connect the DC plug of the 9V-battery holder to the Arduino UNO's DC barrel jack.
8. Connect the power leads from the 4XAA battery box to the power terminal on the Motor Shield.

9. Use double-sided foam tape to attach the ESP8266 module and bi-directional logic level converter. *Figure 7.11* shows the wiring between the Adafruit Motor/Stepper/Servo Shield + Arduino UNO stack and the ESP8266 through the bi-directional logic level converter:

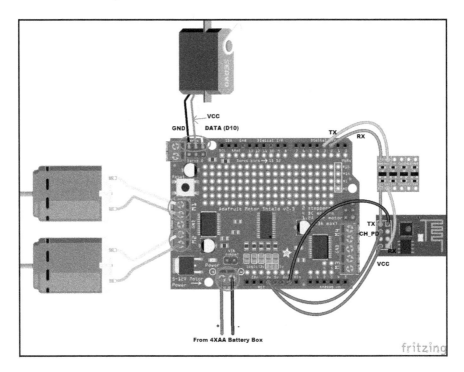

Figure 7.11: Wiring diagram between Arduino UNO and ESP8266 through the bi-directional logic level converter

10. Connect the leads from the servo to the three pin header marked as "Servo 1" on the motor shield. The data line of the three pin header for Servo 1 is internally connected with the Arduino's pin 10.

Controlling the gripper with Blynk

The gripper can be easily controlled with a simple Blynk app. In the previous few chapters, you have written different code for controlling DC motors with Blynk apps. So, this chapter will only focus on how to control the gripper with the Blynk app.

Creating the Blynk app

The following steps will describe you how to add a Slider widget to your Blynk app:

1. Tap anywhere on the canvas to open the widget box.
2. From the widget list, tap on the **SLIDER**.
3. Tap and hold the widget to drag it to a new position if you want.
4. The most important parameter to set is **PIN**. In the **OUTPUT** section, select the output as virtual pin, **V1**.
5. Replace the maximum value, **255** with **1023**.

Software

Listing 7-1 shows the Arduino sketch that can be used to control the gripper with the Blynk app. The functions for controlling the two DC gear motors are not included in the code to only focus on the gripper (servo). Upload the code to the Arduino using the Arduino IDE by connecting the Arduino to your computer with a USB A-to-B cable. After uploading the sketch, remove the USB cable from the Arduino.

Listing 7-1—Controlling the gripper with Blynk:

```
#include <ESP8266_Lib.h>
#include <BlynkSimpleShieldEsp8266.h>
#include <Servo.h>

// You should get Auth Token in the Blynk App.
// Go to the Project Settings (nut icon).
char auth[] = "YourAuthToken";

// Your WiFi credentials.
// Set password to "" for open networks.
char ssid[] = "YourNetworkName";
char pass[] = "YourPassword";

#include <SoftwareSerial.h>
```

```
SoftwareSerial EspSerial(2, 3); // RX, TX

#define ESP8266_BAUD 115200

ESP8266 wifi(&EspSerial);

Servo myservo; //  create servo object to control a servo

// This function will be called every time Slider Widget
// in Blynk app writes values to the Virtual Pin V1
BLYNK_WRITE(V1)
{
  int pinValue = param.asInt(); // assigning incoming value from pin V1 to
a variable. Set min=0 and max=1023 for PIN in your Blynk app.

  // process received value
  int val = analogRead(pinValue); // reads the value of the potentiometer
(value between 0 and 1023)
  val = map(val, 0, 1023, 0, 180); // scale it to use it with the servo
(value between 0 and 180)
  myservo.write(val); // sets the servo position according to the scaled
value
  delay(15);
}

void setup()
{
  // Set ESP8266 baud rate
  EspSerial.begin(ESP8266_BAUD);
  delay(10);

  Blynk.begin(auth, wifi, ssid, pass);

  myservo.attach(10); // attaches the servo on digital pin 10 to the servo
object
}

void loop()
{
  Blynk.run();
}
```

Testing the gripper

It's time to test the gripper with your Blynk app:

1. First, turn on the Mini Rover Robot by sliding the power switch on the 9V-battery box to the **ON** position.
2. Connect the Blynk app with the robot through your Wi-Fi network.
3. If you have configured everything correctly, you can slide the **SLIDER** widget to open and close the gripper. The following are the values for the **CLOSE** and fully **OPEN** states:
    - 0: CLOSE
    - 1023: fully OPEN

Summary

In this chapter, you learned how to build a simple robot with a gripper. Then, you built a Blynk app to control the gripper through Wi-Fi to hold and release lightweight objects.

In Chapter 8, *Photo Rover Robot*, you will learn to build a robot that can be used to take pictures by controlling it through Wi-Fi. The robot is based on the Rover 5 Robot Platform and the camera module is an ArduiCam.

8
Photo Rover Robot

In this chapter, you will build a Rover Robot that can be used to take pictures from remote locations and view them using a web browser. The robot uses a web socket server that allows you to control the camera using a simple web-based interface. The same interface allows you to view the captured image as well. You can save the picture to your computer or mobile device using the web browser for further processing.

In this chapter, you will learn the following topics:

- Building a robot with Rover 5 chassis
- Assembling electronics on Rover 5 chassis
- Connecting ArduCAM with Feather HUZZAH ESP8266
- Creating a web socket server on Feather HUZZAH ESP8266 to serve images to clients

Things you will need

To build the robot, you will need the following things:

- 1 x Dagu Rover 5 tracked chassis (`https://www.pololu.com/product/1550`)
- 1 x Pololu RP5/Rover 5 expansion plate (`https://www.pololu.com/product/1531`)
- 1 x ArduCAM mini camera module shield w/ 2 MP OV2640 for Arduino (`http://www.robotshop.com/en/arducam-mini-camera-module-shield-2-mp-ov2640-arduino.html`) or 1 x ArduCAM mini camera module shield w/ 5 MP OV5642 for Arduino (`http://www.robotshop.com/en/arducam-mini-camera-module-shield-5-mp-ov5642-arduino.html`)

- 1 x FeatherWing Doubler - Prototyping Add-on For All Feather Boards (https://www.adafruit.com/product/2890)
- 1 x Adafruit Feather HUZZAH with ESP8266 Wi-Fi (https://www.adafruit.com/product/2821)
- 1 x DC Motor + Stepper FeatherWing Add-on For All Feather Boards (https://www.adafruit.com/product/2927)
- 1 x Lithium Ion Battery - 3.7V 2000mAh (https://www.adafruit.com/product/2011)
- 6 x Rechargeable NiMH AA batteries: 1.2V, 2200 mAh, 1 cell (https://www.pololu.com/product/1003)
- 1 x Tiny breadboard (https://www.adafruit.com/product/65)

Rover 5 chassis

Rover 5 (*Figure 8.1*) is a tracked chassis that allows you to build robot vehicles that can be driven over many types of surfaces and uneven terrain:

www.pololu.com

Figure 8.1: Rover 5 tracked chassis. Image courtesy of Pololu (https://www.pololu.com)

The chassis consists of the following things:

- Plastic body
- 6-AA battery holder
- Two brushed DC motors

With only four motor leads and two battery leads (*Figure 8.2*), it is easy to interface the chassis with the DC Motor and Stepper FeatherWing:

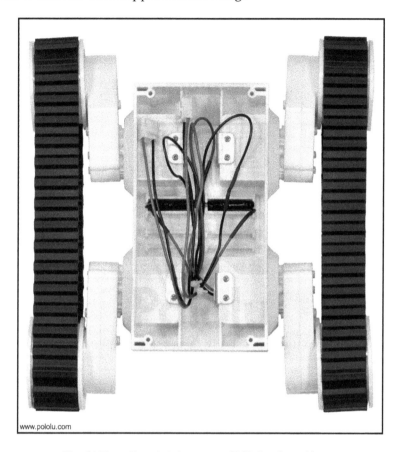

Figure 8.2: Motor and battery leads. Image courtesy of Pololu (https://www.pololu.com)

Additionally, you will need a Rover 5 expansion plate (*Figure 8.3*) (`https://www.pololu.com/product/1531`) to mount all the electronics:

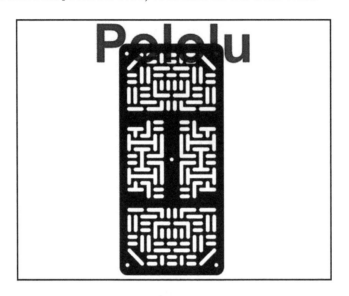

Figure 8.3: Pololu RP5/Rover 5 narrow expansion plate RRC07A. Image courtesy of Pololu (https://www.pololu.com)

Mount the expansion plate on top of the plastic body using the screws provided. Make sure to take all the wires out from the plastic body through the expansion plate before applying the screws.

Connecting the electronics

Let's attach all the electronics to the expansion plate. The following steps will guide you how to attach and wire them:

1. Using four aluminum standoffs, attach the FeatherWing Doubler - Prototyping Add-on to the base plate (*Figure 8.4*):

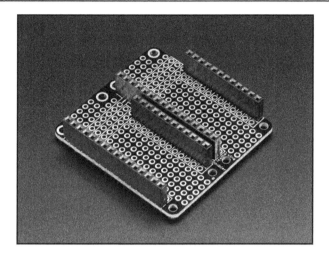

Figure 8.4: Adafruit FeatherWing Doubler - prototyping add-on. Image courtesy of Adafruit Industries (https://www.adafruit.com)

2. Place the Adafruit Feather HUZZAH ESP8266 on the FeatherWing Doubler, as shown in the following figure (*Figure 8.5*):

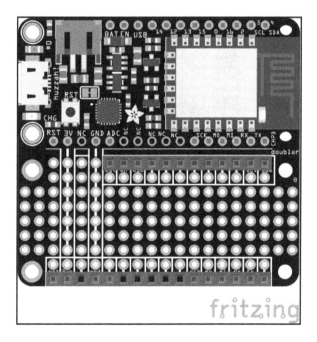

Figure 8.5: Placing the Adafruit Feather HUZZAH ESP8266 on the FeatherWing Doubler

3. Take the Adafruit Stepper and DC Motor FeatherWing and a stick of extra-long male header, and break off five pieces, carefully pulling (or cutting) out the middle one.

4. Insert it into M3 and M4. This will make it easy to connect and disconnect the DC motors (*Figure 8.6*):

Figure 8.6: Adafruit Stepper and DC Motor FeatherWing pre-paired with extra long male headers. Image courtesy of Adafruit Industries (https://www.adafruit.com)

5. Place the Adafruit Stepper and DC Motor FeatherWing on the FeatherWing Doubler next to the Feather HUZZAH ESP8266 (*Figure 8.7*):

Figure 8.7: Placing the DC Motor FeatherWing on the FeatherWing Doubler

6. Connect the wires from the DC motors with the DC Motor FeatherWing. Connect the left motor with the M4 terminal block and the right motor with the M3 terminal block (*Figure 8.8*):

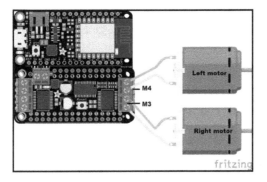

Figure 8.8: Wiring gear motors with DC Motor FeatherWing

7. Connect the + and - lines from the 6-AA battery holder to the DC Motor FeatherWing power terminals (*Figure 8.9*). Don't insert the batteries in the battery holder yet:

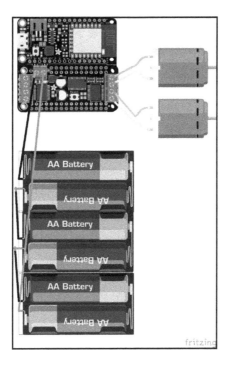

Figure 8.9: Powering DC Motors FeatherWing with a 6AA battery holder

8. Now, attach the ArduCAM on the expansion plate with a small breadboard (*Figure 8.10*), using a piece of double-sided tape:

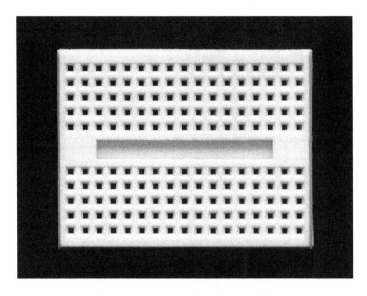

Figure 8.10: Tiny breadboard (https://www.adafruit.com/product/65). Image courtesy of Adafruit Industries

Wiring the ArduCAM with the Feather HUZZAH ESP8266

You can use one of the following versions of the ArduCAM Mini Camera Module Shield (*Figure 8.11*) to connect it with the Feather HUZZAH ESP8266:

- ArduCAM Mini Camera Module Shield w/ 2 MP OV2640 for Arduino
- ArduCAM Mini Camera Module Shield w/ 5 MP OV5642 for Arduino

Figure 8.11: ArduCAM Mini versions

Figure 8.12 shows the wiring diagram between the ArduCAM and the Feather HUZZAH ESP8266:

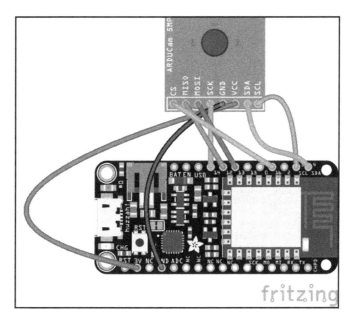

Figure 8.12: Wiring diagram between the ArduCAM and the Feather HUZZAH ESP8266

The wire connections between ArduCAM and Feather HUZZAH ESP8266 are as follows:

- ArduCAM SCL -> Feather HUZZAH ESP8266 SCL
- ArduCAM SDA -> Feather HUZZAH ESP8266 SDA
- ArduCAM VCC -> Feather HUZZAH ESP8266 3V
- ArduCAM GND -> Feather HUZZAH ESP8266 GND
- ArduCAM SCK -> Feather HUZZAH ESP8266 pin 14
- ArduCAM MOSI -> Feather HUZZAH ESP8266 pin 13
- ArduCAM MISO -> Feather HUZZAH ESP8266 pin 12
- ArduCAM CS -> Feather HUZZAH ESP8266 pin 16

Software

This section only focuses on how to capture images with an ArduCAM using a web socket server. You can use any Blynk-based software implementation to control the robot through your Wi-Fi network.

Arduino libraries

You will need the following Arduino libraries to write the Arduino sketch using Arduino IDE:

- ArduCAM
- UTFT4ArduCAM_SPI
- ESP8266-Websocket

You can download these Arduino libraries as a ZIP file (*Figure 8.13*) from `https://github.com/ArduCAM/ArduCAM_ESP8266_UNO`:

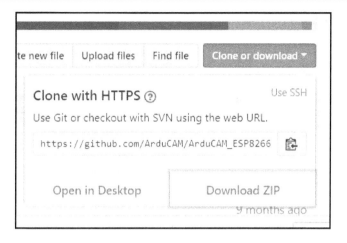

Figure 8.13: Arduino libraries for ArduCAM

The following steps will guide you how to include the previously mentioned libraries to your Arduino IDE and create a web socket server on Feather HUZZAH ESP8266:

1. After downloading the ZIP file, extract it to your hard drive. You will get a folder named `ArduCAM_ESP8266_UNO-master`.

2. Inside the `ArduCAM_ESP8266_UNO-master` folder, you can find a folder named `libraries`.

3. Copy the following folders inside the `libraries` folder to your Arduino IDE's `libraries` folder:
   - ArduCAM
   - UTFT4ArduCAM_SPI
   - ESP8266-Websocket

4. Restart the Arduino IDE to use the newly installed libraries with your code.

5. Then, download the following example Arduino sketch from
   `https://github.com/sumotoy/ArduCAM/blob/master/examples/ESP8266/ArduCA`
   `M_Mini_OV2640_websocket_server/ArduCAM_Mini_OV2640_websocket_server.in`
   `o` (*Figure 8.14*):

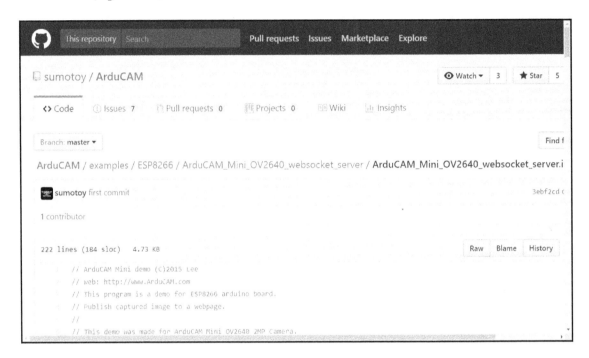

Figure 8.14: ArduCAM_Mini_OV2640_websocket_server.ino example Arduino sketch

6. After downloading the example Arduino sketch, open it using the Arduino IDE.
7. Connect the Feather HUZZAH ESP8266 with your computer using a micro-USB cable.
8. Make sure that the following Arduino libraries are included in your Arduino IDE's `libraries` folder:

```
#include <ESP8266WiFi.h>
#include <WiFiClient.h>
#include <ESP8266WebServer.h>
#include <ESP8266mDNS.h>
#include <Wire.h>
#include <SPI.h>
#include <ArduCAM.h>
#include <WebSocketServer.h>
```

9. Change the following lines according to your Wi-Fi network's configuration settings:

```
const char* ssid = "your sid"; //This is your Wi-Fi router's
name const char* password = "your sid pass"; //This is your Wi-
Fi routers password
```

10. After configuring the Arduino sketch, flash it to the Feather HUZZAH ESP8266. Then, remove the micro-USB cable from the Feather HUZZAH ESP8266 and connect the 3.7V 2000 mAh Li-poly battery to the battery connector of the Feather HUZZ10AH ESP8266.

11. Insert 6 AA batteries in the battery holder on the Rover 5 chassis.

12. Then, open the `camera_demo.html` file located in your `examples\ESP8266\ArduCAM_Mini_OV2640_websocket_server\html` folder (*Figure 8.15*):

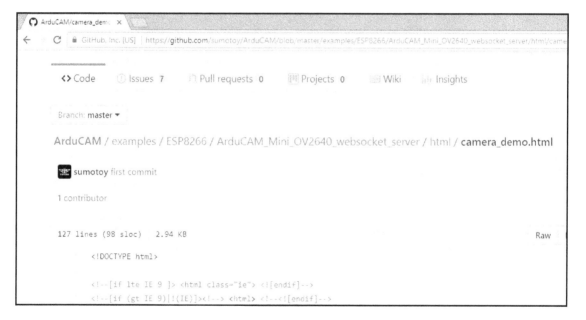

Figure 8.15: camera_demo.html

13. Type the IP address of the web socket marked in the red box as shown in the following picture (*Figure 8.16*). You can find the IP address of your Feather HUZZAH ESP8266 with your Wi-Fi router's admin portal. You can also assign a dynamic or static IP address to ESP8266 with your router:

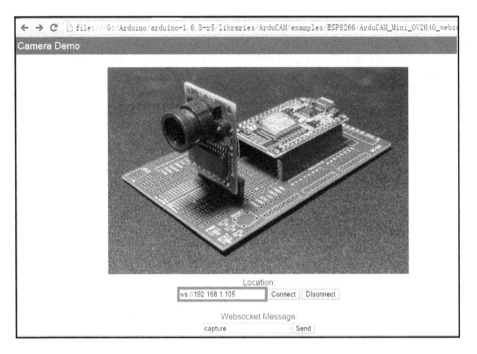

Figure 8.16: Inserting server URL

14. Click on the **Connect** button; you will see the websocket connection request from the serial monitor. After a few seconds, the webpage will notify the user that a successful connection has been established.

15. Then, click on the **Send** button to send a websocket message to the ESP8266. It will trigger the capture command on the ESP8266 and send back the captured image and display it on the webpage. (*Figure 8.17*):

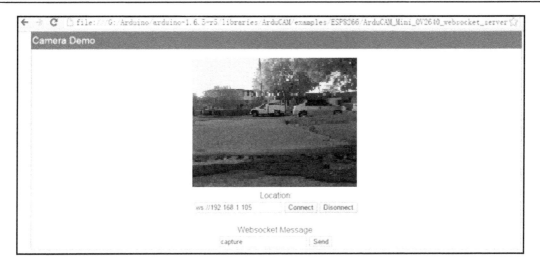

Figure 8.17: Captured image

16. Using the web browser, you can save the captured image to your computer for further analysis.

Summary

In this chapter, you learned how to capture an image using an ArduCAM and how to send it to the user's browser using an ESP8266-based web server through a Wi-Fi network. You also learned how to control the Rover 5 robot using a simple web page with control buttons.

Throughout this book, you learned about different types of applications that can be built with an ESP8266. You used variations of ESP8266-based modules such as the original ESP8266 (ESP-01) and the Feather HUZZAH ESP8266. However, you can build the same projects using other ESP8266 modules available in the market with minimum modifications of both hardware and software. You can further enhance any project presented in this book with your own creativity for use in real-world applications.

Index